T0093550

Network Evolution and Applications

Network Evolution and Applications provides a comprehensive, integrative, and easy approach to understanding the technologies, concepts, and milestones in the history of networking. It provides an overview of different aspects involved in the networking arena that includes the core technologies that are essential for communication and important in our day-to-day life. It throws some light on certain past networking concepts and technologies that have been revolutionary in the history of science and technology and have been highly impactful. It expands on various concepts like Artificial Intelligence, Software Defined Networking, Cloud Computing, and Internet of Things, which are very popular at present.

This book focuses on the evolutions made in the world of networking. One can't imagine the world without the Internet today; with the Internet and the present-day networking, distance doesn't matter at all. The COVID-19 pandemic has resulted in a tough time worldwide, with global lockdown, locked homes, empty streets, stores without consumers, and offices with no or fewer staff. Thanks to the modern digital networks, the culture of work from home (WFH) or working remotely with the network/Internet connection has come to the fore, with even school and university classes going online. Although WFH is not new, the COVID-19 pandemic has given it a new look, and industries are now willfully exploring WFH to extend it in the future. The aim of this book is to present the timeline of networking to show the developments made and the milestones that were achieved due to these developments.

Network Evolution and Applications

Vikas Kumar Jha, Bishwajeet Pandey, and
Ciro Rodriguez Rodriguez

CRC Press is an imprint of the
Taylor & Francis Group, an **informa** business

First edition published 2023
by CRC Press
6000 Broken Sound Parkway NW, Suite 300, Boca Raton, FL 33487-2742

and by CRC Press
4 Park Square, Milton Park, Abingdon, Oxon, OX14 4RN

CRC Press is an imprint of Taylor & Francis Group, LLC

ISBN: 9781032299563 (hbk)
ISBN: 9781032299549 (pbk)
ISBN: 9781003302902 (ebk)

DOI: 10.1201/9781003302902

Typeset in Times
by codeMantra

Contents

Preface

Networking is one of the glorious technological evolutions in the history of mankind. Communication being one of the essential utilities to humans, network evolution plays an important role in making it more consumable. Communication is one of the essential requirements of our world and, more specifically, the society we live in. It simply means the exchange of information or the transfer of views, ideas, and feelings from one to another, from one to many. Communication has some key requirements such as two or more parties to communicate, a medium or a communication channel, a common language, and the information. The setup of the platform through which effective communication can happen is part of the communication technology that also includes a network setup. Communication technology has changed a lot with time and so has changed the network that has impacted our lifestyle. This book focuses on the evolution in the world of networking. One can't imagine the world without the Internet today; with the Internet and the present-day networking, distance doesn't matter at all. The COVID-19 pandemic has resulted in a tough time worldwide, with global lockdown, locked homes, empty streets, stores without consumers, and offices with no or fewer staffs. Thanks to the modern digital networks, the culture of work from home (WFH) or working remotely with the network/ Internet connection has come to the fore, with even school and university classes going online. Although WFH is not new, the COVID-19 pandemic has given it a new look, and industries are now willfully exploring WFH to extend it in the future.

The aim of this book is to highlight the important aspects of networking and to represent the overall timeline of networking with the developments made and their impact on our life. This book presents the journey of Network Evolution: its past—history and success of the project ARPANET; its present—the modern-age networking where the SDN and NFV are making networking software-centric and customizable at the software level; and its future—where AI will play a major role in making the network more agile and collaborative. This book highlights the important concepts, terminologies, and innovations in networking. This book also discusses how innovations in networking have impacted the human life and the digital disruption of the present age.

We still have certain milestones to achieve in the field of networking. Networking has a bright future ahead, and its evolution is a never-ending one. We all are somewhere a part of the network and are somewhere connected. This book is an effort to establish a connection, a network, with its readers.

Authors

Vikas Kumar Jha earned his MTech in computer science engineering from ABV—IIITM, Gwalior, with specialization in advanced networks and a BE degree in electronics and communication engineering from RGPV Bhopal. He did his MTech thesis on Quantum Communication Networks and has five international journal publications under his name. His areas of research include communications, networks, cloud computing, and AI in telecommunication. He has received the following global certifications: Cisco Certifications (CCNA—Routing and Switching and CCNA—Security), Juniper Certification—JNCIA, Amazon Certification—AWS Certified Solution Architect Associate, and Microsoft Certification—Azure Cloud Fundamentals. He has more than 8 years of experience in telecommunication, including with Tata Communications Limited, formerly Videsh Sanchar Nigam Limited. He has also taught undergraduate engineering students as a lecturer in an engineering college for a year.

Prof Dr Bishwajeet Pandey earned his PhD in computer science engineering from Gran Sasso Science Institute, L'Aquila, Italy, under the guidance of Prof Paolo Prinetto, Politecnico DiTorino (World Ranking 13 in Electrical Engineering). He has worked as an assistant professor in the Department of Research, Chitkara University; Junior Research Fellow (JRF) in South Asian University; and lecturer at the Indira Gandhi National Open University. He completed a Master of Computer Applications (MCA) and a Master of Technology (VLSI) from ABVIIITM Gwalior along with R&D project from CDAC-Noida. He is an associate Professor at the Department of Computer Science and Engineering, Jain University, Bangalore, India. He has authored and coauthored 137 papers available on his Scopus Profile: https://www.scopus.com/authid/detail.uri?authorId=57203239026. He has 1400+ citations according to his Google Scholar Profile: https://scholar.google.co.in/citations?user=UZ_8yAMAAAAJ&hl=en. He has experience in the teaching of Innovation and Startup, Computer Network, Digital Logic, Logic Synthesis, and System Verilog. His areas of research interest are green computing, high-performance computing, cyber-physical systems, artificial intelligence, machine learning, and cybersecurity. He is on the board of directors of many startups of his students, e.g., Gyancity Research Consultancy Pvt Ltd.

Prof. Dr. Ciro Rodriguez Rodriguez is a professor-researcher at the National Universities Mayor de San Marcos and Federico Villarreal. He completed his PhD in engineering, did advanced studies at the Institute of Theoretical Physics (ICTP), Italy, and in the United States Particle Accelerator School (USPAS), and studied information technology development policy studies, Korea Telecom (KT), South Korea. His areas of research interest include artificial intelligence, health-social welfare, and environment. He holds two patents in the Patent Office INDECOPI in Peru.

1 Communication Network at a Glance

ABBREVIATIONS

ARIB	Association of Radio Industries and Businesses
ATIS	Alliance for Telecommunications Industry Solutions
BS	Base station
BTS	Base transceiver system
CCITT	Consultative Committee for International Telephony and Telegraphy
CCSA	China Communications Standards Association
CT	Core Network and Terminals
ETSI	European Telecommunications Standards Institute
IANA	Internet Assigned Numbers Authority
ICANN	Internet Corporation for Assigned Names and Numbers
ICT	Information communications technology
IETF	The Internet Engineering Task Force
ITU-T	International Telecommunication Union – Telecommunication
LAN	Local area network
MAN	Metropolitan area network
MTSO	Mobile telecommunication switching office
OTT	Over-the-top
PAN	Personal area network
PSTN	Public switched telephone network
RAN	Radio access network
RF	Radio frequency
SA	Services and systems aspects
TCP/IP	Transmission Control Protocol/Internet Protocol
TSDSI	Telecommunications Standards Development Society of India
TTA	Telecommunications Technology Association
TTC	Telecommunication Technology Committee
UE	User equipment
WAN	Wide area network
WWW	World Wide Web
3GPP	3rd Generation Partnership Project

1.1 INTRODUCTION

A computer network is one of the exciting and interesting fields of communication. In the simplest way, it can be considered as a group of computers or communicating nodes (devices like servers, other networking hardware, printers, and so on) that use

DOI: 10.1201/9781003302902-1

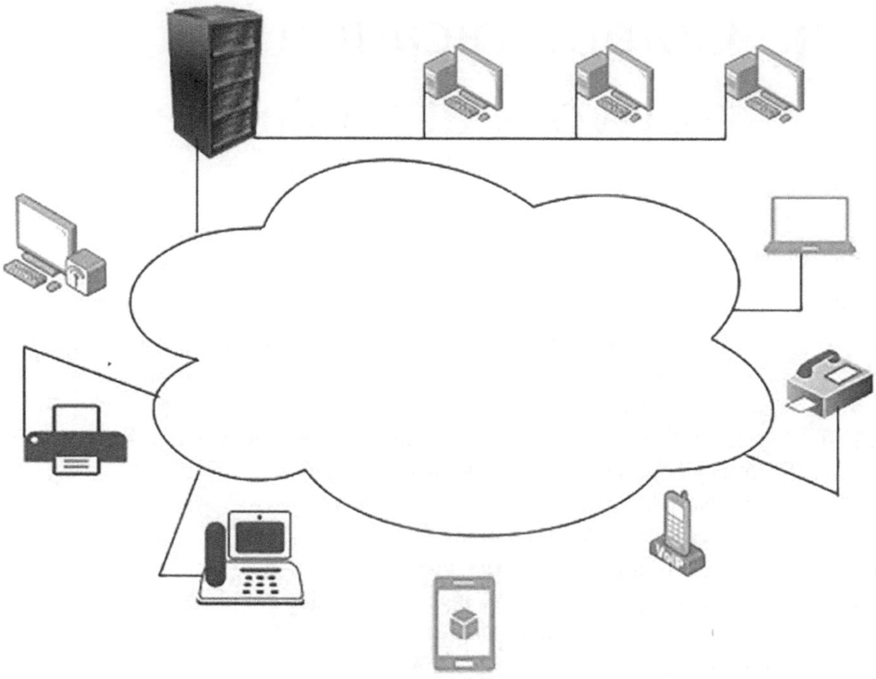

FIGURE 1.1 Diagram to represent communication network around us.

a set of common communication protocols over some interconnections for the goal of sharing resources among themselves. The communication among these connected nodes takes place in the form of electronic signals. There is always a need of the establishment of a network between those computers or communicating nodes with a common set of communicating protocols to help establish the exchange of data. The Internet on the other hand can be thought of as a biggest communicating global network with millions of communicating nodes globally. The use case of a computer network is always increasing and is not limited to sending emails, World Wide Web, transfer of files, sending a print or facsimile request, multimedia, over-the-top contents, and many more (Figure 1.1).

1.2 TYPES OF NETWORKS

There are a variety of networks around us having their broad range of classification. Every network type in use has its own utility, is based on certain technology or standard, and has some parameters and scope that make it useful. Networks are composed of several devices that perform certain network functions and follow a layered architecture that we will discuss more in the next chapter. As Figure 1.2 depicts, the classification of the network is based on several parameters that we are going to look at.

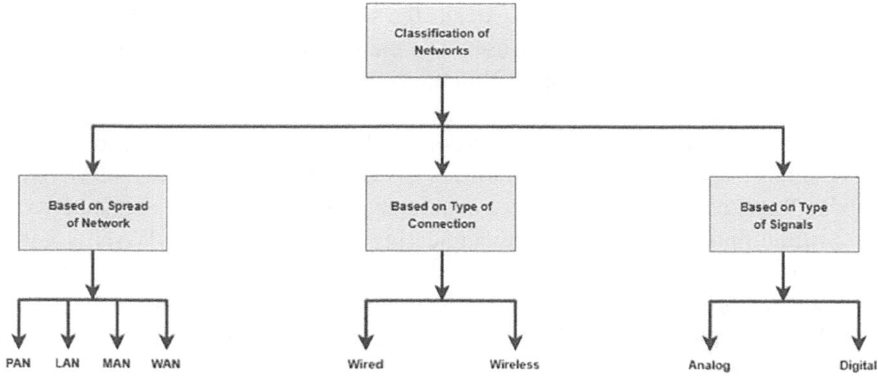

FIGURE 1.2 Classification of network.

1.2.1 Based on Spread of the Network

Every network has a certain geographical coverage, and the devices connected within that coverage area can take part in communication effectively. So, classification on the basis of the spread has its own place in the domain of networking. Classification of networks on the basis of spread of the network or the geographical distance that a network cover has been done as a personal area network (PAN) (a network with the spread limited to a personal area or room), local area network (LAN) (a network with the spread within a building), metropolitan area network (MAN) (a network with the spread within a metropolitan area or inside a city), and wide area network (WAN) (a network extending the MAN capability and connecting two or many cities).

PAN: A network that facilitates the communication or the exchange of data between the devices within the vicinity of someone's personal space is called as Personal Area Network or PAN. So, the coverage will be limited to a small area like a room, personal workspace or an office cabin, and so on. They can be used for communication among computers, smartphones, gadgets, or other electronic devices, and they are also sometimes connected with the Internet by having the PAN connection to the Internet gateway. Many of the PAN technologies are getting widely adopted these days to provide connectivity for the Internet of things implementation. The PAN can be of the following types: the wired PAN (e.g., USB, FireWire) or wireless PAN (e.g., Bluetooth, Infrared, ZigBee).

LAN: A network that supports the communication among interconnected computers at one physical location such as residence, building, office, school, Laboratory, and university campus is called as local area network or LAN. A typical LAN can support the connection of few computers up to thousands of computers or devices; however, the only thing that characterizes a LAN is its scope within a limited area. A LAN can have a wired network (Ethernet) or a wireless network (Wi-Fi).

MAN: A computer network that extends the scope of a LAN and establishes a connection to incorporate the users of a metropolitan area is the MAN. The MAN is a common network within the geography of a city that can support multiple LANs.

WAN: A WAN is the further extension of the scope of a MAN, and it can support a very large number of users as compared to the LAN and MAN. A WAN can be considered as a computer network providing connectivity to multiple MANs or for interconnecting multiple cities. A WAN is not location specific as compared with the LAN and MAN, and it has a much wider presence, considering the example of the Internet as the biggest WAN of the globe. A WAN network is often provided by the telecommunication service provider and is further leased to the business, users or public. Corporates having international offices located in the different parts of the world often lease WAN connectivity from the service providers for establishing a network connectivity between their offices located globally.

1.2.2 Based on the Type of Connection

On the basis of the connecting media type, networks have been broadly classified as a wired network (a network having a fixed line cable or physical connection between connecting nodes) and a wireless network (a network with no proper line established or the media is not a physical cable).

- A wired network uses physical cables to transfer data between the devices connected through the network. There are different types of cables available which support this network such as copper wire, twisted pair, or fiber optic. Wired connectivity has been the conventional means of networking, and it is a general assumption that a wired network can provide high security with high bandwidth provisioned for each user. Wired connectivity is considered

TABLE 1.1
Comparison of Wired and Wireless Networks

Factors	Wired Network	Wireless Network
Communication medium	Physical cable	Air
Standard	IEEE 802.3	IEEE 802.11
Mobility	Limited	High
Security	High	Low
Speed	High (up to 1 Gbps)	Lower than wired network
Network access	Physical access	Proximity access
Delay	Low	High
Reliability	High	Low
Installation cost and time	High	Low
Equipment	Router, switch, hubs, and so on	Wireless router, access points, and so on

as a highly reliable network that incurs very low delay comparative to a wireless network.

- A wireless network [1], on the other hand, can be seen as a network that does not require any physical cable to establish the network as well as to connect the user devices with the network. Wireless network uses the electromagnetic waves like infrared or radio frequency signals to transfer data between devices. This allows devices to stay connected to the network while also roaming freely without being tethered to any wires. Wireless networks are also known as Wi-Fi which is very popular nowadays. We can often find a Wi-Fi hotspot around us when we go to a café, hotel, airport, railway station, hospital, and many more public places, and at these places, we are provided with the facility to connect wirelessly our devices through the hotspot's wireless network.

1.2.3 Based on the Type of Signals

Networks classified based on the communicating signals are of the following types: the analog network – uses analog technology to transfer signals and the communicating signals are of analog nature – and digital network – uses digital technology and the communicating signals are digital in nature. A signal is an electromagnetic or electrical current that is used to carry the information from one device in the network to another. Analog and digital are the types of transmission signals, and hence, these are mostly related at the level of signal transmission of the network. In a network, there are many of the devices which can understand analog signals only or the digital signals only. In that case, the general approach while designing a network is to use the conversion mechanism to convert the signals from analog to digital and vice versa whenever required in the network. Figure 1.3 represents the waveform of analog and digital signals.

1.3 NETWORK TOPOLOGY

Computer networks have different network topologies [2–4] which represent the way of establishing the connection and the transfer of the data among them. These topologies are very useful when we are designing a network, specifically, a LAN setup as per our requirement.

Few popular local area network topologies are mentioned below:

- Bus topology
- Star topology

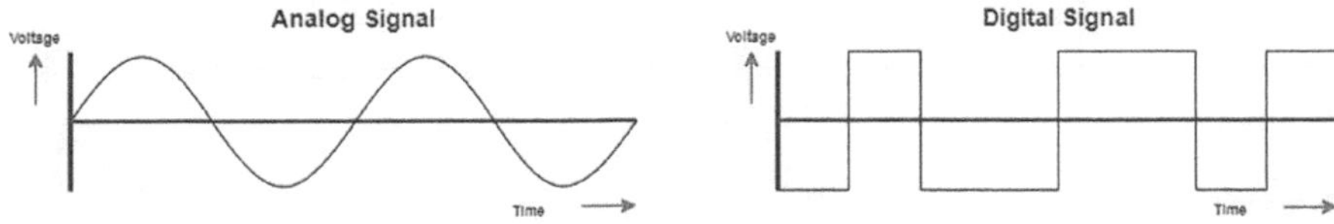

FIGURE 1.3 Waveform of analog and digital signal.

- Mesh topology
- Ring topology
- Hybrid topology

1.3.1 Bus Topology

Bus topology uses a single main cable to connect multiple nodes, so a single cable will be used for data communication among the connecting nodes. This main cable connecting all the nodes is also sometimes called as backbone cable. Figure 1.4 represents an example of a bus topology network.

Advantages:

- Simple connection and easy network installation.
- Require less cable length for the network setup.

Disadvantages:

- Not a scalable network topology, useful only for small networks.
- Use of a single cable can create segments of the network and can be difficult in fault isolation.

1.3.2 Star Topology

Star network topology uses a device centrally connecting all the other nodes individually. This central device is also called as the hub of the network. Two nodes can only communicate through the hub. The network architecture is also sometimes called as the hub-spoke architecture. Figure 1.5 represents a typical star topology diagram.

Advantages:

- Easy connection as all the other devices are connected to the hub.
- Easy fault isolation due to the separate connection of every node.

Disadvantages:

- The hub is the single point of failure, and failure of the hub will create the entire network down, and every node will be isolated with another node.

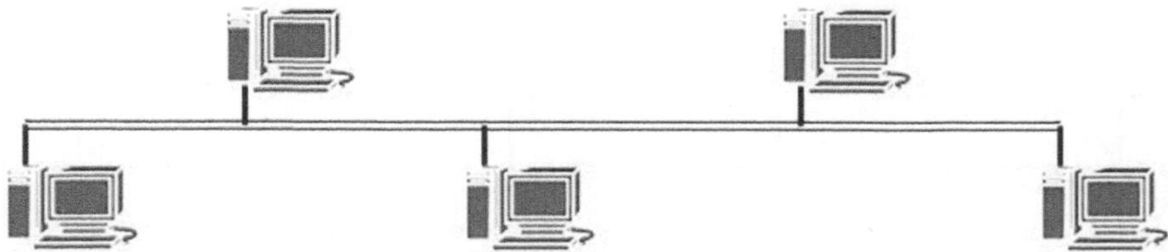

FIGURE 1.4 Bus network topology diagram.

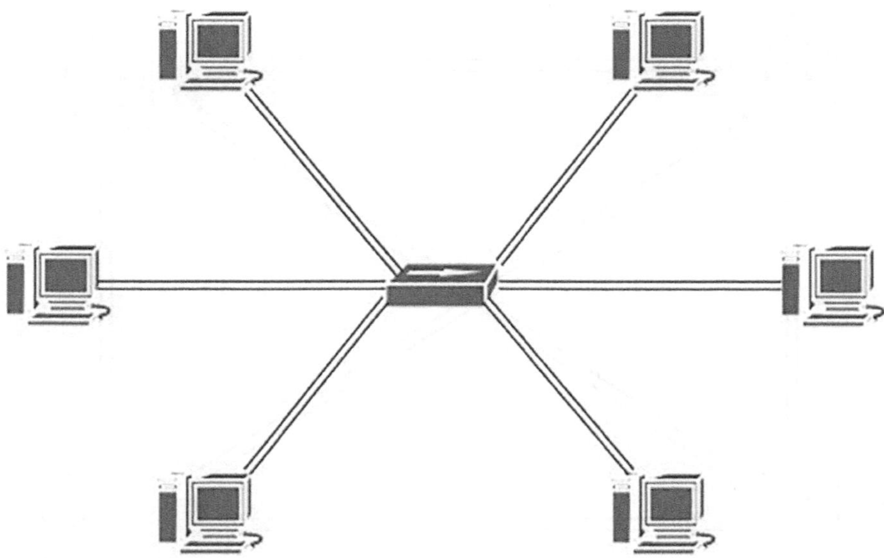

FIGURE 1.5 Star network topology diagram.

1.3.3 Mesh Topology

Mesh topology is that network topology which has the most complex cabling connection as every node is to connect every other node with a dedicated cable connection. The data communication between any of the two nodes will happen directly due to the dedicated cable connecting them. If a network in mesh topology has n number of nodes, then the number of cables required for the mesh network will be $n(n-1)/2$. Figure 1.6 represents a typical mesh topology diagram.

Advantages:

- Data communication is fast with less traffic due to a direct dedicated cable between communicating nodes.
- Easy fault isolation of the mesh network.

Disadvantages:

- The most complex cable network architecture.
- Mesh network is not scalable due to increasing cable connection.

1.3.4 Ring Topology

Ring topology has the network formation of a ring by connecting each node with exactly two nodes connected directly on the either side of it. Data communication among the nodes follows a circular data path and crossing over each of the intermediary node between the communicating nodes. Figure 1.7 represents a typical ring network diagram.

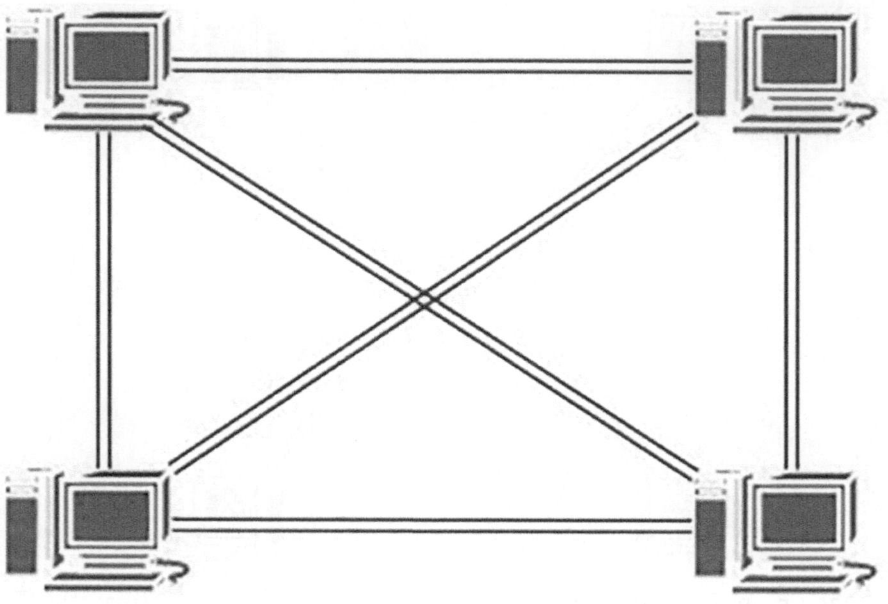

FIGURE 1.6 Mesh network topology diagram.

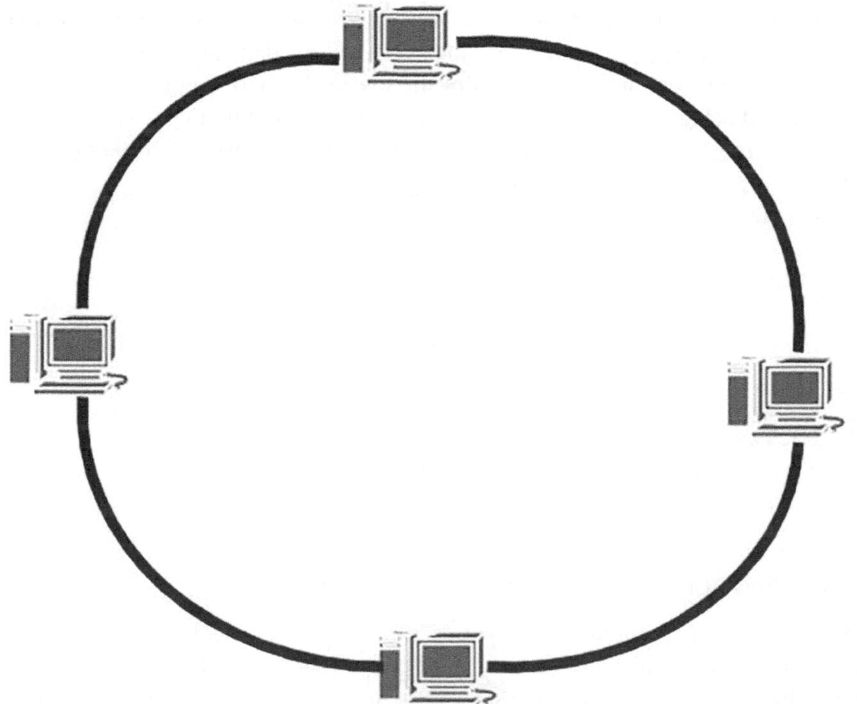

FIGURE 1.7 Ring network topology diagram.

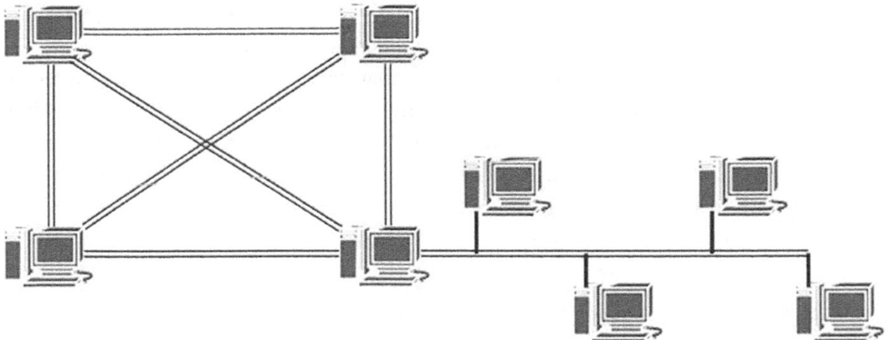

FIGURE 1.8 Hybrid network topology diagram.

Advantages:

- Reduced packet collision as the flow of data is in one direction.
- Easy to install network.

Disadvantages:

- Failure of a single node can disrupt data communication in the entire network.
- Slower network and high data traffic as the data packet needs to pass through several nodes.

1.3.5 Hybrid Topology

Hybrid topology is a network topology that can be formed by using any of the two network topologies mentioned above. For example, we can have a hybrid topology by connecting a bus network with a star topology or a ring topology. Figure 1.8 represents an example of hybrid network.

Advantage:

- Network topology can be more suitable based on requirement.
- Scalable network topology.

Disadvantage:

- Most complex network design due to combination of network topologies.
- The network setup can be costly.

1.4 DATA COMMUNICATION

Data communication, in general, is the process of transferring data from the sending node toward the receiving node. It happens only after a network connection is established between the communicating nodes. The terms data communication and computer network are very much related and dependent on each other. The data received by the receiving node should ideally replicate the data that was generated and transmitted by the sending node. This data can be any information or a message signal that can be digital or analog that the sender wants to send it to the receiver. Usually, there are certain data losses in the communicating medium, and the received signal may not represent the exact data as was transmitted by the sending node, and hence, there are certain techniques of error correction mechanism to get corrected data at the receiving end.

1.4.1 Use Cases of Data Communication

There are numerous and uncountable use cases of data communication in our daily life as it is so much related with our day-to-day activities. Some of the use cases have been listed below:

Emails: Emails have got much importance in our daily life. The term electronic mail became popular as "email" with the intention of emulating the physical letter to be presented digitally and can be shared through computer systems over a network. Enterprises are using business emails very commonly and provide it to their employees to have business communications among themselves and with the customers. Email communication also provides the feature to attach several documents that is again a very useful use case of data communication.

Instant Messaging: It is another popular use case of the data communication in present that provides real-time messaging facility over the Internet, also popular as Instant Messaging technology. Examples are the chat services from several social apps available like Facebook messenger, WhatsApp, WeChat, Viber, and so on. Many of the enterprises are providing Instant Messaging based platforms to the employees for the communication among the employees.

Voice Communication: Voice calls are the most common and essential use case of the data communication that has now multiple technologies, platform, and media available to support. It provides the capability of speaking with someone through technology supporting audio communication in the real time.

Video Communication: Video calling and video conferencing are other very common and important use cases of communication, and technology innovations have worked to make it more and more convenient, cheaper, and useful.

Text Messaging or Short Messaging Service: Short messaging service is yet an important use case of data communication and is one of the oldest and widely used text messaging technologies.

Facsimile Communication: Facsimile or Fax is the communication method that uses telephone lines to transmit the scanned image of the printed document that is further to be reproduced at the receiving end by connecting the output of the telephone lines to a printer.

1.4.2 Different Data Communication Mode

As we discussed about the term "data communication" in our earlier section, we can now clearly say that the process of data communication will involve data transmission and data reception through the communication channel, where the data can be either in the form of the digital signal or analog signal. The purpose of data communication is not successful until both of the individual activities, the data transmission and the data reception, are complete.

During the process of communication, the devices or the communicating nodes have the ability to either transmit only or receive only or both. Based on this ability, the flow of data communication has been divided into three communication mode:

Simplex: It is that data communication mode in which the sender device can only transmit data and not receive, and the receiver device will only receive data and not transmit. The communication is unidirectional in nature, where we have fixed sender and receiver devices. Considering a channel or link with two communicating parties at its end (a point-to-point link), we can think of it as one of them will be the sender and the other will be the receiver and both will utilize the full channel link.

Half Duplex: Considering the example of a point-to-point link with two parties, each connected at the end of the channel or link. In the half duplex mode, the communicating parties have the ability to both transmit and receive, but they can perform only one activity at a time. So, the flow of data will only be unidirectional at a time, but data flow can happen in both directions. Comparing simplex and half duplex, we can find that in the simplex mode, the sender can only send and will not expect any response back from the receiver; however, in the half duplex mode, the sender will transmit the message and can also expect a response back from the receiver. One device on a point-to-point link has to wait until the data transmission from the other device is complete.

Full Duplex: A full duplex data communication mode has the devices with the ability to transmit and receive data at the same time. A full duplex mode provides a two-way or bidirectional communication, and the channel can be considered as composed of two unidirectional simplex channels in both the directions. Therefore, both of the devices on a point-to-point link will be transmitting and receiving message from the link.

1.4.3 Components of Data Communication

We have now seen the importance of data communications in our daily life and are familiar about the concept of a communication network. We will see several components of data communications.

Sender: A sender is that communicating device or node which wants to send the data or message signal through the data communication channel to the receiving node.

Receiver: A receiver is another communicating device or node that will receive the data or message sent by the sending node.

Message: A message is the data or signal that is sent by the sender and received by the receiver. For an effective data communication to happen, the message sent by the sender and received by the receiver should be same.

Channel or Link: A sender and receiver have a communication channel, which is also called as link, established between them. Until the link is not established between both the nodes, the data communication will not be possible. A link can be established by using wired or wireless communication media.

Protocols: During communication between the nodes of a communication network, the nodes are supposed to follow a set of standard rules or communication guidelines that is also called as the communication protocol.

Modulator: Modulation is a technique used to influence the message signal or the data to be transmitted using another signal called as the carrier signal. This carrier signal can be seen as the signal which is used to carry the message signal with it through the communication channel. There are certain limitations of using a message signal alone while communicating, and to have a good quality of message received at the other end, it is supposed to transmit the modulated signal. A modulator is a device that is used to module the message signal.

Demodulator: We know about the modulation technique used at the time of transmitting the signal; now, we can simply say that the received signal by the receiver is not the actual message signal from the sender. Correctly, the signal received is the modulated signal, not the actual message signal from the sender. So now, at the receiving end, some other type of technique is to be used to recover the exact message that was sent by the sender. This process of getting the message signal from the modulated signal is the inversion of the modulation process and hence called as the demodulation process. A demodulator is that device which is used to do this activity.

Modem: A modem is that device which can modulate and demodulate signals; it is the single device that performs both of these activities. The purpose of a modem is to translate the digital signal coming from a computer into an analog signal for its transmission over the communication or telephone lines. Similarly, it can translate the incoming analog signal from the communication lines into a digital signal that is to be understood by the computer receiving that signal.

Multiplexer: Talking about the communication network in general, we can think of it as a network with multiple devices trying to communicate with each other. A general requirement was that there will be multiple sender and receiver nodes at a time who will be trying to send signals through the communication channel, and we need some technique that will make this possible with the same communication channel supporting message signals from multiple sources to be transmitted through it without delay. Therefore, multiplexing was a technology that was introduced to combine message signals from multiple sources in a network and that signal now called as multiplexed signal is to transmit over the network. The device that performs this activity is called as multiplexer.

Demultiplexer: Demultiplexing is the inverse process of multiplexing; the receiving node should get the same message signal that it was supposed to get and the same has been sent by the source node. Hence, the demultiplexing technique is used to separate out all the individual message signals that were part of the multiplexed signal which will be separately received by individual receivers. The device that can perform this activity is known as demultiplexer.

Repeater: Communication networks often suffer with a problem of the signals getting lost or distorted due to the certain limitations of the communication media which forces a low-quality signal to be received at the receiving end. Hence, there is a requirement of regenerating the same signal that was transmitted after a certain distance. To accomplish this task, the repeater was introduced in the communication network. These repeaters also help to extend the network coverage.

Router: Routing is an important concept and phenomena of a communication network. Routing can be seen as the process of guiding a data packet to consider the best possible path or the best route to reach the destination node. Routers are those devices which take routing decision within the network, and to facilitate this, a router is needed to store all the routing path information or route record entry in its database. Routers are configured with a certain route selection procedure known as the "routing protocol" based on which the router identifies the best route to the destination network. A network consists of multiple devices like we discussed earlier, and it also has routers. Routers are the Layer 3 (network layer) devices of the Open System Interconnection (OSI) model and Transmission Control Protocol/Internet Protocol layers.

Switch: Switches are those devices which are used to guide the data frames to reach the destination host in the communication network from the available switched path; this process is called as switching. Routers and switches are used to form and extend a network; and hence, these are essential components. Switches are usually the Layer 2 (data link layer) devices of the OSI model.

1.4.4 Cellular or Mobile Communication

Cellular communication [5], also called as mobile communication, is an important achievement of the history of telecommunication that is based on the cellular radio network. A wireless communication technique uses radio frequency signal carriers, and the network is distributed geographically with spatially distributed cells of certain area, providing network coverage to the mobile phone. Each cell in the network consists of a base transceiver system also known as a base station (BS) within itself working in certain frequency defined that provides radio coverage to the user equipment like mobile phone handsets, tablets, and so on. Cells are defined with different carrier frequencies usually with the neighboring cells to avoid the interference.

- Each cell is equipped with a low power transmitter as part of its BS and is allocated with a range of frequencies that can support a number of subscribers.
- A BS is approximated to the center of each cell and includes an antenna, a controller along with a number of transceivers for communicating on the channels that were assigned to the cell.
- The frequencies allocated to one cell can be reused or assigned to some other cell, except the neighboring cell, in the network so as the cochannel interference or crosstalk can be avoided.
- Important design decision during the formation of the cellular network is the consideration of the cell size and cell radius that will provide good radio coverage up to its boundary; hexagonal cells are usually convenient.
- The BS of each cell is connected with one mobile telecommunication switching office (MTSO), where a single MTSO can serve multiple BSs. These MTSOs facilitate in order to establish connection between different user equipment.
- The MTSO is also having connection with the public switched telephone network, and therefore, it can make connection between a public switched telephone network customer or fixed subscriber to mobile subscriber.
- The handoff mechanism is an important feature of cellular or mobile communication where a call can be successfully handoff from one cell to another with no loss or interruption to call during the movement of the mobile equipment.

1.4.5 Internet

The Internet is the largest communication network worldwide that spreads across our globe and provides a cheapest communication means to us. It is the Internet technology that has revolutionized the communication industry and has a very significant impact in the digitalization of the present world [6]. The Internet has the potential to touch many aspects of our life multifold and can be accessible to every individual of the society. The Internet is such a giant platform that has taught us collaboration of several organizations, collaboration of different networks, collaboration of different technologies, coexistence of several vendors, and so on. The Internet has reshaped or redefined most traditional communication media including telephony, radio, television, newspapers, and many more. The birth of social media, online music, online video, streaming media or over-the-top platforms, online books, online studies/classrooms, websites, e-marketing/e-business, and so on has proved us the importance of the Internet in our daily life.

1.5 INTERNATIONAL FORUMS AND ORGANIZATIONS

1.5.1 ITU-T

Telecommunication Standardization Sector of the International Telecommunications Union (ITU-T) [7] formerly known as the Consultative Committee for International

Telephony and Telegraphy is an international organization that was established to standardize and regulate international radio and telecommunications. It has a study group that assembles experts from around the world to develop international standards known as ITU-T Recommendations which further act as defining of the global infrastructure of information and communication technologies (ICTs). The development of the standards is critical for the purpose of interoperability of ICTs whether there is the exchange of voice, video, or data messages. Standards enable global communications by ensuring that countries' ICT networks and devices are speaking the same language. Standardization also avoids costly market battles over preferred ICT. They create a level playing field which provides access to new markets for the companies from emerging markets. ITU-T has driven from its inception in 1865 as a contribution-led, consensus-based approach to develop standards in which all countries and companies, no matter how large or small, are afforded equal rights to influence the development of ITU-T Recommendations. It became a specialized agency of the United Nations in 1947 with its headquarters in Geneva, Switzerland, next to the main United Nations campus.

1.5.2 IANA

The Internet Assigned Numbers Authority (IANA) [8, 9] is an administrative function of the Internet which is responsible for coordinating some of the key elements that keep the Internet running smoothly. The IANA has taken the coordination role for the Internet as there was always a technical need for some key parts of the Internet to be globally coordinated because of the reason that the Internet is a renowned worldwide network free from any central coordination. The IANA allocates and maintains unique codes and numbering systems that are used in the technical standards ("protocols") to drive the Internet. Various activities of the IANA are grouped into the categories as domain names, number resources, and product assignments. So, basically, we can say that the IANA is that organization which keeps track of global Internet protocol addresses and their allocation, autonomous system number allocation, domain names management of root zones, and Internet protocol parameter identifiers that are used by Internet standards. It is one of the oldest institutions on the Internet, with the IANA functions dating back to the 1970s. The IANA was made as an operating unit of the Internet Corporation for Assigned Names and Numbers, an American multistakeholder group and nonprofit organization responsible for coordinating the maintenance and procedures of several databases related to the namespaces and numerical spaces of the Internet.

1.5.3 IETF

The Internet Engineering Task Force (IETF) [10] is an open standards organization or a large open international community of the network designers, operators, vendors, and researchers concerned with the evolution of the Internet architecture and the smooth operation of the Internet. The IETF develops and promotes voluntary Internet standards and, in particular, the standards that comprise the Internet protocol suite (transmission control protocol/Internet protocol) with all of its participants and managers as volunteers;

however, their work is usually funded by their employers or sponsors. It follows an open process so that any interested person can participate in the IETF work, can know what is being decided, and make the voice to be heard on the issue. The technical work of the IETF is done in its working groups, which are organized by topic into several areas including routing, transport, security, and so on. The IETF makes their documents, Working Group mailing lists, attendance lists, and the meeting minutes publicly available on the Internet. The IETF produces its documents for the technical competence needed and is also willing to listen to the technically competent input from any of the source to have the "engineering quality" in its design. The IETF makes the standards and also takes the ownership of all the aspects of a protocol standard based on the combined engineering judgment of its participants or volunteers and the real-world experience in implementing and deploying of those specifications.

1.5.4 3GPP

The 3rd Generation Partnership Project (3GPP) is an umbrella term for a number of standards organizations and unites seven of the telecommunications standard development organizations which are ARIB, ATIS, CCSA, ETSI, TSDSI, TTA, and TTC; those are known as "organizational partners". It has seven of the organizations national or regional telecommunication standards organizations mentioned above as primary members and also has a variety of other organizations as associate members. It provides a stable environment to the members to produce the reports and specifications that define 3GPP technologies according to Ref. [11]. Projects covered by the 3GPP include cellular telecommunications technologies including radio access, core network, and service capabilities, which provide a complete system description for mobile telecommunications. It also provides interface for nonradio access to the core network and for interworking with non-3GPP networks. 3GPP provides the specifications and studies which are contribution-driven by their member companies, in working groups and at the technical specification group level. Three technical specification groups of the 3GPP are radio access networks, services and systems aspects, and core network and terminals.

1.6 CONCLUSION

Communication is a human need, and technology innovations have successfully worked on its growth with continuous evolutions. Establishing communication by connecting two separate and isolated parts of the globe through a network is possible because of the efforts made in this direction. In this chapter, we have tried to explain the basic concepts, terms, and technology for the understanding of a computer network.

REFERENCES

1. Kolluru, D. S., & Reddy, P. B. (2021). Review on communication technologies in telecommunications from conventional telephones to smart phones. In *AIP Conference Proceedings* (Vol. 2407, No. 1, p. 020003). Hyderabad, India: AIP Publishing LLC.

2. Forouzan, B. A. (2021). *Data Communication and Networking*, 6th edition, New York: McGraw Hill.
3. Tanenbaum, A. S., Feamster, N., Wetherall, D. J. (2021). *Computer Networks*, 6th edition, New York: Pearson.
4. Connectivity: Wireless & Wired, Electronics Notes. Accessed: Feb. 5, 2021. [online]. Available: https://www.electronics-notes.com/articles/connectivity/
5. Stent, G. S. (1972). Cellular communication. *Scientific American*, 227(3), 42–51.
6. Castells, M. (2014). The impact of the internet on society: A global perspective. *Change*, 19, 127–148.
7. CCITT -50 Years of Excellence, International Telecommunication Union. Accessed: Feb. 6, 2021. [online]. Available: https://www.itu.int/ITU-T/50/docs/ITU-T_50.pdf
8. IANA, IETF. Accessed: Feb. 8, 2021. [online]. Available: https://www.ietf.org/standards/iana/
9. IANA Functions: The Basics, ICANN. Accessed: Feb. 8, 2021. [online]. Available: https://www.icann.org/en/system/files/files/functions-basics-07apr14-en.pdf
10. Alvestrand, H., A mission statement for the IETF, RFC 3935. doi:10.17487/RFC3935, October 2004, [online]. Available: https://www.rfc-editor.org/info/rfc3935
11. Flynn, K., About 3GPP, 3GPP. Accessed: Feb. 8, 2021. [online]. Available: https://www.3gpp.org/about-3gpp

2 Reference Model and Protocol Suite

ABBREVIATIONS

ARPANET	Advanced Research Projects Agency Network
DARPA	Defense Advanced Research Projects Agency
DoD	Department of Defense
FTP	File Transfer Protocol
HTTP	Hyper Text Transfer Protocol
IPv4	Internet Protocol Version 4
IPv6	Internet Protocol Version 6
ISO	International Standard Organization
LLC	Logical link control
MAC	Media access control
NIC	Network interface card
OSI	Open Systems Interconnection
PDU	Protocol data unit
PPDU	Presentation PDU
QoS	Quality of service
SMTP	Simple Mail Transfer Protocol
SPDU	Session PDU
TCP/IP	Transmission Control Protocol/Internet Protocol
TPDU	Transport PDU
UDP	User Datagram Protocol

2.1 INTRODUCTION

We have gone through the concepts of the computer network in the previous chapter, and now, we have some background of communication. It is time to explore more about the communication network in terms of the protocol. This chapter will give you the understanding of the Open Systems Interconnection (OSI) reference model and the Transmission Control Protocol/Internet Protocol (TCP/IP) suite that are important pillars of the computer network [1]. We should have a good grasp over the protocol architecture for understanding of a computer network. We will see the motivations behind this layered approach of protocols. We will know about different layers of the computer network, and we will be able to understand the relation between them and how the layered approach is supporting communication.

DOI: 10.1201/9781003302902-2

2.2 WHY STANDARD PROTOCOL ARCHITECTURE?

Why there was a need of protocol standardization in communication? A simple question that comes in our mind when we start our discussion about the standard protocol architecture in network. Well, a reference model was always a requirement to support multiparty, multivendor communication. A network having multiple devices from different manufacturers can communicate easily if they follow the common approach to handle data and process it. So, it should have certain architecture for the specific tasks that are to be performed by each of the individual component. Before doing that, it was a prime concern to divide the computer network into certain components (software or hardware) and associate each of the component with certain tasks to perform. The performing of task will follow some specified guideline where every component has some predefined instruction of when to perform that task and for whom. The components that we just discussed are the layers of communication model, and the predefined tasks are the protocols associated to each layer. This was the birth of the layered protocol architecture that became vital in communication and produced many recommendations in the journeys ahead. The layered architecture was composed as a vertical stack of the component layers, where each component layer in the stack has to perform the task to support the communication with the other system [2]. From there, the concept of peer-to-peer data communication was introduced and that became an integral part of the protocol architecture. Each layer has to offer some processed data output of its function to the layer above.

During the initial phase of networking, it was observed that with the growth of the computer network, there were so many implementations of it. Each of the implementation was working independently and using its own protocol architecture or language to communicate with other systems. These protocol architectures and their network implementation were different from each other. Different organizations were having their own distinct network that could not communicate to each other. The device manufacturer has to follow any of the implementation for the hardware and software for the device and has to accept that at the cost of not compatible with other network implementations. This pushed the communication industry to search for a reference standard of the protocol architecture. The invention of the TCP/IP [3] was the milestone that solved many problems. It set a common framework of the layered protocol architecture so that the independent network implementations can have a common intermediary protocol that would allow them to communicate with each other. Suddenly, it became very popular as different organizations showed interest and started working with following TCP/IP. It gave a way to establish communication among distinct networks and is considered as a breakthrough and a de facto standard among the networking community. It was this breakthrough which further gave us the Internet, the largest collaboration of networks overall.

2.3 LOGISTICS OF COMMUNICATION

Before starting our discussion on the reference protocol architecture, let us discuss about the logistics of communication. We should learn some of the key requirements from a computer network, some of the important parameters which will make the communication successful. We will know more about the structure of the supposed to be the reference model. Therefore, this section has been termed here as the logistics of communication. The reference model that we will learn in the next section will include them.

2.3.1 Protocols

A protocol is a set of rules that has a significant place in international relations. It is considered as an agreement between several nations where protocols minimize the problems caused due to the cultural differences between them. This provides a common understanding with a set of guidelines that everyone follows and interprets the actions of others. Similarly, in data communication, when multiple devices have to communicate, it is evident to define some kind of set of rules. This rule will work as the guideline to govern the communication of the devices. This set of rules which is very popular in networking is the data communication protocols.

2.3.2 Layers

We discussed in the previous section about the need and motivation of the layered protocol architecture. The idea of creating layers in the communication system is to create a logical separation of the overall network.

Layers are the division of the complete network function into multiple subfunctions that are to be performed individually to make the communication successful. Each layer has an important role to play and is helpful in reducing the complexity overall if we consider the communication as a single network function.

We can understand that a computer network will have a number of computers that will communicate to each other through the network. The computers or end nodes will be used for either sending or receiving data, by that way fulfilling our goals from the network. We can refer to the network's application as the user's goals that are achieved by the end nodes. The computers will be running those applications and are providing an interface user to interact. Considering this, we can name the layer which is doing all the task related to interacting with the user and supporting network application as the application layer. The application layer will be the higher most layer of the computer network and will hold data as per interaction from the user.

The connection of the different devices in the network will be considered as another layer which will be used to transfer the electrical or electromagnetic waves/signals through them. This layer's function should be the transfer of signals in its most elementary form from one point to another. This layer will behave as the carrier of information and can be called as a physical layer. The physical layer will be the lowermost layer which will hold the network data and transfer them to the desired destination in the form of signals.

We need one application layer which will be interfacing with the user's application on the one end and will be supporting the exchange of application data through the network.

There is a requirement of some type of control over the users' application and need to keep the application data from different applications separately.

We need some type of a data converting mechanism so that the users' understandable data will be converted into a certain suitable format which can easily be transferred through the network.

2.3.3 Service

Services and protocols are distinct concepts of the computer network. A service is a set of operations that a layer provides to the layer above it in the layer stack. The service defines the operations that a layer is prepared to perform on behalf of its users and says nothing about how these operations are implemented. A service is related to an interface between two layers; the lower layer can be called as the service provider and the upper as the service user. And therefore, we can say that the layered architecture in the reference model that we have in data communication is also somewhat a type of the service-oriented architecture. The service-oriented architecture is an architectural concept that focuses on different services communicating with each other to carry out a bigger job.

Service and the protocol are completely decoupled with each other, where services relate to the interfaces between layers and protocols relate to the implementation of their service definitions as such is not visible to the intended user of the service. In contrast, a service defines operations that can be performed but does not specify how these operations are implemented.

2.3.4 Client/Server Method

The client/server method of communication represents a technique in which one of the systems/machines is considered as a client and another system/machine is the server [4]. Both can be considered as two different computers or systems which want to communicate. It is always the client that will put some request for service and that request will be served by the server. There is the need of a network connection between the client and server. A client/server model of communication has generally multiple clients and a centralized server (Figure 2.1).

The request for service from a client will be based on the user's application. Consider one machine that has some data, call it a server, and another machine remotely connected with this machine by a network where the user is sitting, call this machine a client. Now, the user on the client machine wants to have that data which are currently present on the server, so the user will put a request for data from the client machine to the server machine. The server will check the request, identify the client machine that has requested for the data, and then will send the data. Each time the client wants that data, it has to send a request to the server. This way of interaction between the systems on the network is called as the client/server communication model.

2.3.5 Addressing

To reach the exact destination, we should have the correct address. If we have to send a courier or a letter by post, we are supposed to mention the correct address on top of it. That address will include certain specific details like the name of the person (the intended receiver of the letter), house number, street name or number, city, post/zip

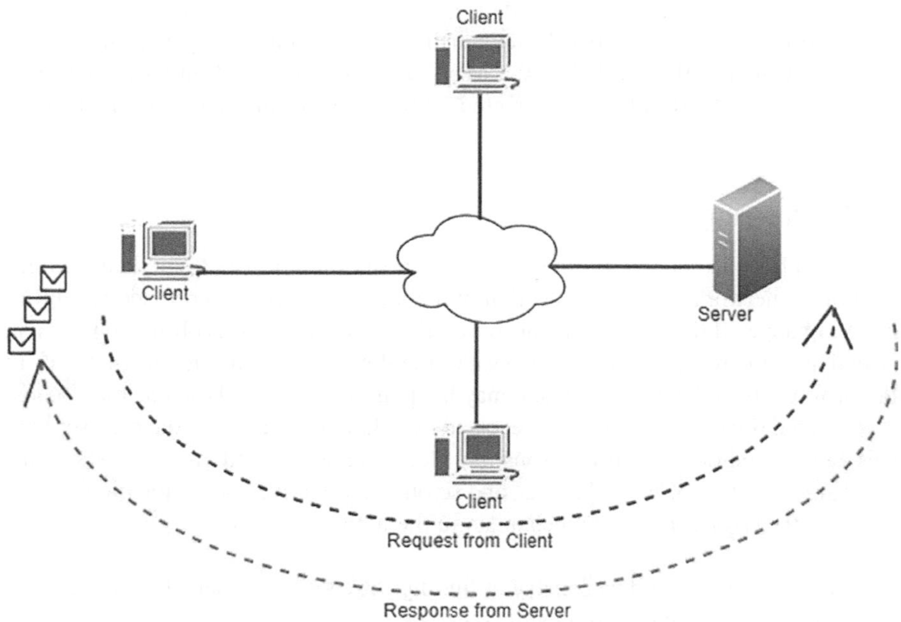

FIGURE 2.1 Client-server communication.

code, country, and so on. If the address is correct and the same has been followed in the transport of the letter, then there are good chances that it will be successfully delivered to the intended receiver. On the similar way, the address of the destination device or a network can be used in data communication to guide the data packet to reach the correct destination (the intended receiver) through the network. Since the network may have a number of computers connected, it will need a mechanism to identify the senders and receivers of a particular message. Hence, addresses are very important for a successful communication, and the computer network should need some addressing mechanism to support it.

In the computer network, we have two broad categories of address which are very common. These are the logical address and the physical address.

Logical Addressing: Logical addresses are the addresses which are assigned to a network or a system by the network administrator or the Internet service provider. A logical addressing method is such a technique of addressing where a logical address lets you access a network device by using an address that you assign. The logical address is associated with the network layer of the reference model, and the Internet Protocol (IP) address is the popular example of the logical address.

Physical Addressing: A physical address is the hardware address which is embedded on the network interface card of the computer. Physical addressing is the hardware addressing mechanism of the network's device, and it is usually assigned by the manufacturer of the hardware.

The media access control (MAC) address is the example of physical addressing. Every network device that we buy has a physical address associated with it called as the MAC address imprinted on its interface card. This MAC address imprinted on the network interface card is fixed and cannot be changed.

2.3.6 Reliability

The reliability of a network is often one of the most desirable aspects of data communication. A network having a collection of devices is supposed to provide reliability that will happen if the devices operate correctly and no data loss is observed in transit as well as at the reception of data. Consider the data transfer through the network in the form of bits in the data packets; it may happen that some of these bits may be lost or corrupted (inverted) and will be received as a damaged packet at the receiver end. This can happen due to many reasons like electrical noise, random wireless signals, hardware- or software-related issues, and so on. An effective communication system will have the error correction mechanism in the network inbuilt.

- The network should be capable of finding the errors and then allow the corrected data to be received by the receiver.
- One such method is the **error detection** technique that will allow the detection of the error whenever the information is received incorrectly. The information which has been received incorrectly can then be retransmitted until it is received correctly.
- Another method available to provide the data reliability is the **error correction** technique that will allow the correct message to recover from the possibly incorrect bits that were originally received.
- Both methods use addition of redundant bits along with the message bits that will allow the detection or correction of the error at the receiver end.
- Another design issue regarding reliability is to find the best suitable path to reach the intended receiver from the many paths available in the network. This will help in the reduction of network delay or latency of the received packet. The network is supposed to take a decision to choose the best network route; this concept is called as **routing**.

2.3.7 Flow Control Mechanism

A **flow control** mechanism is used to tell the sender that how much data should be sent so as the same will be received and processed by the receiver. The purpose is to control the flow of data with no data loss caused by the sender sending the data at a speed higher than that at which the receiver is able to receive and process it. The flow control method is used in networking to ensure that the sender and receiver are both synchronized in terms of the data rate, preventing data loss. This mechanism allows an acknowledgment from the receiver before sending the next data until which the sender has to wait.

There are two popular protocols used for the flow control:

1. Stop and wait protocol
2. Sliding window protocol

Stop and Wait Protocol: Every time the sender will send one data frame, it will stop and wait for an acknowledgment from the receiver. It will send next data frame only after the acknowledgment has been received for the last data frame. It will send the next frame and will stop and wait again for an acknowledgment from the receiver. This process will continue until the entire data frames have been received with their acknowledgments from the receiver.

If it has been observed that any of the frame was lost and not received by the receiver, then the receiver will not send any acknowledgment as it has not received that frame. The sender will wait for an acknowledgment until a specific time as the ***time-out*** timer, and after the time-out expires, it will repeat the last frame for which there was no acknowledgment from the receiver. Figure 2.2 represents the frame sequence for stop and wait protocol.

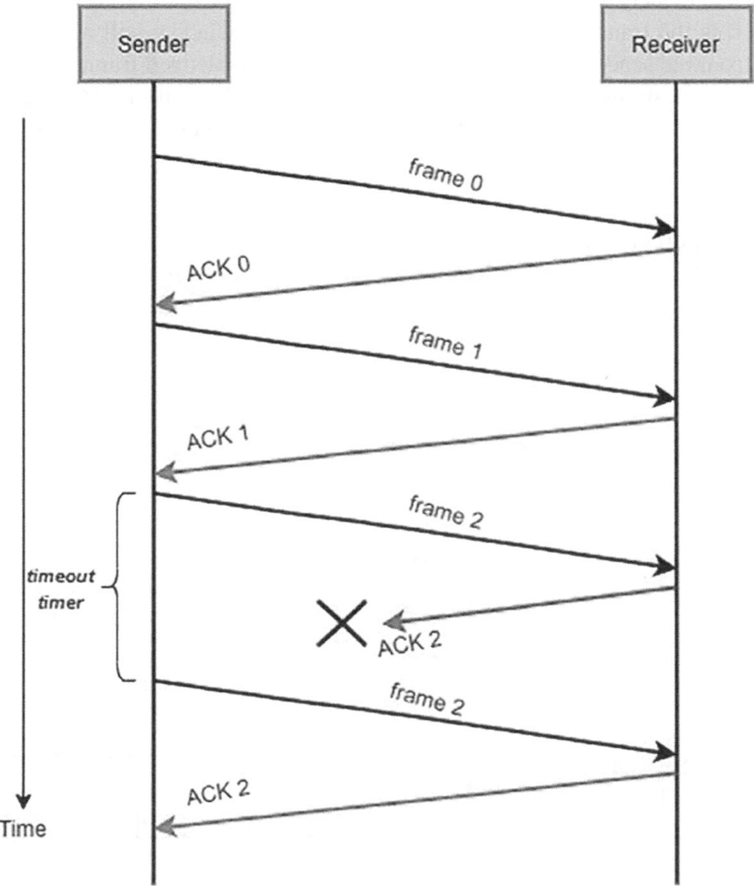

FIGURE 2.2 Stop and wait protocol.

Advantage:

- Implementation in the network is simple.

Disadvantage:

- Only one data frame can be sent at a time.
- Efficiency would be low if the distance between the sender and receiver is large.
- Data transfer through the network would be slow as with each frame, the sender has to stop and wait for acknowledgment.

Sliding Window Protocol: A sliding window protocol was introduced to reduce the data transfer delay from the network as compared to the stop and wait protocol. It is using a windowing technique, where a window can be thought of as a buffer collection of "n" numbers of frames. All the frames in the window are numbered from 0 to n–1. A sender will send n frames at a time and will keep the record of n frames in a window. A receiver will send acknowledgment of the frames that have been received. As the acknowledgment of the frame is received, its record is to be removed from the window, considering the frame that was already received. This window will now have a new frame from the sender along with the previous unacknowledged frames. By this way, the window is sliding to the next frame with respect to the unacknowledged frames. Figure 2.3 represents the frame transfer sequence for the sliding window protocol.

Advantages:

- Better efficiency and network throughput.
- Sliding window method do not waste network bandwidth.

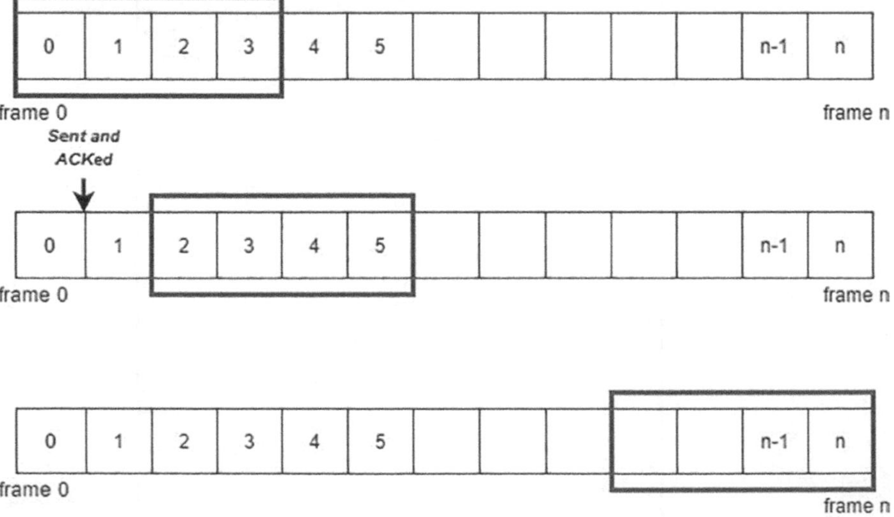

FIGURE 2.3 Sliding window protocol.

Disadvantages:

- The estimation of the suitable size of the window is quite complex.

2.3.8 Connection Oriented – Connectionless

This is an important distinction between the protocols which offer the transmission of the data through the network. Sending of the data through the network can follow either a connection-oriented service or a connectionless service.

Connection Oriented: A connection-oriented service is designed and developed after the telephone system. The connection-oriented protocol requires a logical connection to be established between the two systems before data is exchanged through the network. This connection should be maintained during the entire duration of communication and then be released afterward. The process was developed considering the telephone system, and it is much like a telephone call, where a logical connection as a virtual circuit is established during the entire duration of the call. The caller must know the person's telephone number, pick up the phone, dial the number, talk, and then hang up the call. In the earlier telephone system, the telephone networks have a circuit-based connection established over a copper wire that carried a phone conversation. The orders in which the data is delivered remains same in the connection-oriented service. The Transmission Control Protocol (TCP) is the example of the connection-oriented service. A connection-oriented service is more reliable than the connectionless service.

Connectionless: A connectionless service is designed and developed after the postal services and allows the data to be exchanged without establishing a link between two systems within a network. Each of the messages (letters) carries the complete destination address and is sent independently to the destination without considering any specific order of subsequent messages. The connectionless service follows a similar approach where the messages (packets) from the source can follow the path through the intermediate nodes inside the network independently of all the subsequent messages. The order in which the messages are sent and the order in which messages are received may be different. The User Datagram Protocol (UDP) is an example of the connectionless service.

2.4 TCP/IP

The TCP/IP [5,6], which we just discussed as a breakthrough in protocol architecture of the layered protocol stack, refers to an entire suite of data communication protocols. The TCP/IP architecture is the outcome of the research and development work on the experimental packet-switched network that was funded by the Defense Advanced Research Projects Agency. It happened after the U.S. Department of Defense sponsored the Advanced Research Projects Agency Network, a research network that connected hundreds of universities and government installations, using leased telephone lines. We will discuss in detail about the Advanced Research Projects Agency Network in next chapters.

The TCP/IP reference model gets its name from two of the very important protocols that belong to it: the TCP and the IP. The protocols of the TCP/IP model have been so popular in networking that they have been widely accepted and used across industries rather than the TCP/IP as a complete reference model. This is the reason that the TCP/IP is widely known as the TCP/IP protocol suite than the TCP/IP reference model. For the consideration of the standard reference model, there is another model that has the recipe of the definitions of a suitable layered protocol architecture, the OSI reference model; we will discuss more about the OSI model in detail in the next section of this chapter.

2.4.1 Layered Architecture

The TCP/IP model has the layered stack of five layers which are the physical layer, link layer, Internet layer, transport layer, and application layer. Sometimes, the physical and link layer are combinedly represented as the network access layer [7]. These layers are the logical segmentation, not the physical segmentation, of the computer network based on the tasks that are assigned to a particular layer. Each layer is encapsulated with protocols which fulfill the layer's functionality (Figure 2.4).

Physical Layer: It is the lowest layer of the TCP/IP model which covers the physical interface between a data transmission device and transmission medium within the network and specifies the characteristics of the network hardware. The physical layer deals with data in the form of signals/bits and handles the host-to-host data transfer in the network. This layer defines the medium of transmission and communication mode between two devices. The medium of data transmission can be wired or wireless, and the communication mode can be simplex, half duplex, or full duplex.

Link Layer: The link layer is the second layer of the TCP/IP model layered stack over the physical layer. The link layer has dependency on the physical layer as it deals directly with the transmission medium. So, the link layer

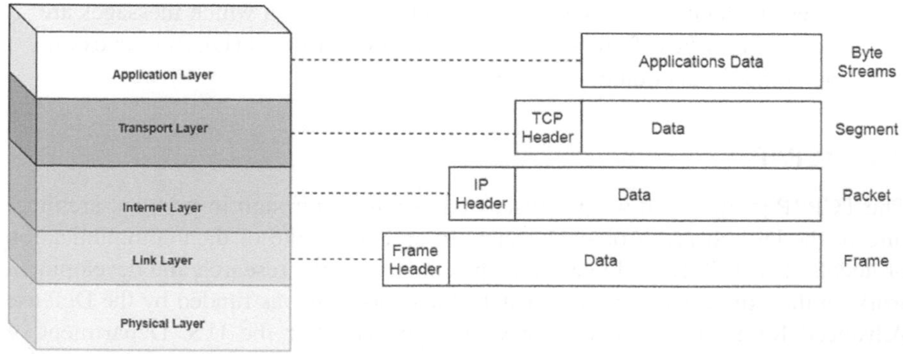

FIGURE 2.4 TCP/IP model layered architecture.

has to always support the data coming from higher layers to be translated and converted into data frames that will be then interacting with the physical layer (the transmission medium) to transfer the data in the form of signals to the destination device. Data framing is an important process at the link layer in which it has to add some header information to the data packets coming from the higher layer for the successful delivery of data packets to correct destinations. To achieve this, the link layer performs physical addressing of the data packets by adding the source device's and the destination device's physical address to it. This process is also called as the data encapsulation by the link layer.

Internet Layer: This layer is also called as the network layer. It has protocols which support the data communication between the different networks, and therefore, it facilitates with the capability of internetworking. Since the Internet is the combination of multiple networks, it would require a certain set of protocols which will support the transfer of data packets to the correct network. The most powerful protocol supporting the Internet layer is called as the IP, which is a connectionless protocol. That means, this protocol provides no guarantee to the data packets which are sent or received in the correct order, along the same path, or even in their entirety. The IP is supporting the routing of IP packets at this layer to provide the routing function across interconnected networks. A router is a processor that connects two or multiple networks and has its primary function to relay the IP packet from one network to another on its route from the source to the destination. The Internet layer also has a companion protocol called as the Internet Control Message Protocol that helps it function.

From the earlier layers, we can think of them as the data is being communicated between the network and the devices, so they are mostly supporting us to access the data from the link and provide it to the correct destination host. For that, the physical addressing is used by the link layer while it is encapsulating the packet into data frames. However, the Internet layer–based protocols are using the concept of logical addressing (IP addressing) and not the physical addressing to encapsulate the data from the higher layer into IP packets. This encapsulation will use the IP address of the destination network in the header. There are two versions of the IP: IPv4 and IPv6.

Transport Layer: It is the fourth layer of the TCP/IP model above the Internet layer and fulfills the role of end-to-end transporting of the application's data from one process to another successfully. The transport layer deals with data coming from a higher layer as the applications data and then transports it as data segments to its peer destination host. To perform this, the transport layer has to create segmentation of the received data from the higher layers at the source system. After transportation of the segmented data to the peer system, a reverse-segmentation activity is to perform at the transport layer of the destination host. This activity is called as reassembly of segments to covert the actual applications data. Therefore, **segmentation** and **reassembly** are the two important tasks for the transport layer that will also depend on whether the data segments are free from errors when received at the destination host. That constitutes the requirement of third task for the transport layer which is **error control**.

The transport layer has two protocols: the TCP and UDP. The TCP provides a reliable transport of data segments and is connection-oriented; hence, it will always require establishing transport layer connection before sending of data. The TCP also performs sequencing of data segments, flow control, and error control in data transmission. However, the UDP is a connectionless protocol and does not perform sequencing, flow control, and error control in data transmission. The UDP does not provide acknowledgment for the received data and is an unreliable protocol. It is fast as compared to the TCP and is suitable for real-time data items, client-server-type request-response queries, and applications which require prompt delivery than accurate delivery.

Application Layer: It is the upper layer of the TCP/IP model that provides an interface between the network services and the application programs executing on the host machines. The application layer provides services over the network to the end users and contains the logic needed to support various applications. The application layer synchronizes the application data at the source host and at the destination host. For each of the user's application, a separate module specific to that application is needed.

Various application layer protocols are the File Transfer Protocol (FTP), Telnet, Simple Mail Transfer Protocol, Hyper Text Transfer Protocol (HTTP), and so on.

2.4.2 TCP/IP Operation

The TCP/IP model has the client-server method of communication where a user or client machine requests for a service. The response is provided by another machine considered as a server that provides a service like sending a webpage through the network to the client machine. The protocol suite of the TCP/IP has been classified as stateless protocols, which means each of the client requests is treated as a new request as it is unrelated to the previous one. However, the transport layer has the two protocols, the TCP and UDP, in which the TCP is stateful and UDP is stateless protocol. The TCP is a connection-oriented protocol, so it is stateful. As it has to maintain the connection until the end-to-end delivery of the applications, data is successful.

The TCP/IP model is somewhat different with the OSI model that was designed after the TCP/IP.

Figure 2.5 shows the TCP/IP layered protocol architecture to represent the end-to-end communication between Host A and Host B. The overall network will be composed of interconnecting multiple networks as we can see in the diagram. To access resources or services from the network, the connected host through the network will be using some network access protocol; the Ethernet methodologies are the most common example. To establish communication between one network and another, we need to use the routing methodologies. Routers will be used to connect its interfaces through cable connection with other networks. The configuration of the

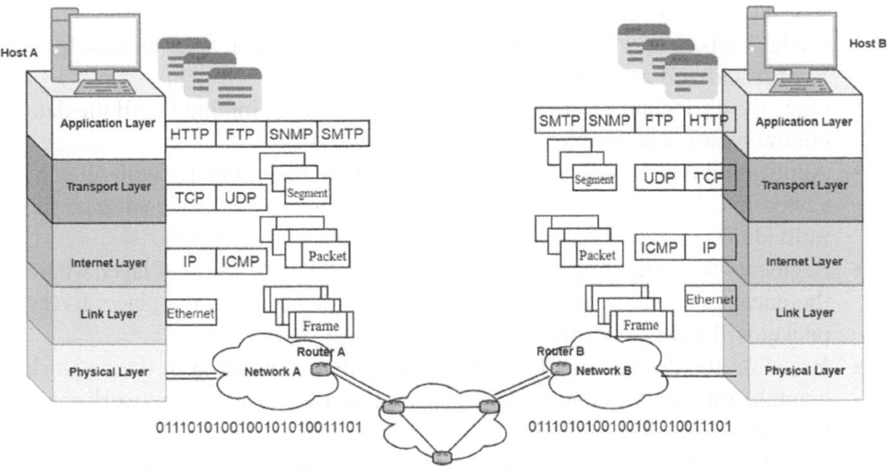

FIGURE 2.5 TCP/IP model.

router will be required to use any of the routing protocol. Routers will take a routing decision based on the configured routing protocol when a packet will arrive at its any of the interface to route the IP packet to the destination network. The end-to-end communication has been shown in the diagram and can be explained in below points.

- Host A can be considered as connected with Access Network A based on an Ethernet connection and, similarly, Host B with Access Network B.
- The Networks A and B have no direct connection and are connected through Network C in between them.
- We can assume that Host A and Host B want to communicate with each other based on any of the TCP/IP model application layer protocol like FTP, Simple Mail Transfer Protocol, HTTP, and so on.
- Every application layer protocol is given a unique port number for its identification by the lower layer, like FTP – 21, Simple Mail Transfer Protocol – 25, and HTTP – 8080.
- Both host machines can be considered either as a client or a server and so the user on one host can request for service from the other host.
- Let's assume the user on Host A will send a request for service to Host B.
- Host A should know the destination IP address, i.e., the IP address of Host B. We have understood earlier that the logical address to identify a network or a host connected to a network is done by IP addressing.
- Host A also needs the application port number and will use that application port number to communicate with Host B. The combination of the IP address with the application port number is called as a socket.
- Host A will send the TCP connection request to Host B using the Host B IP address and destination port number.
- The network should be reachable between Host A and Host B, and the TCP connection needs to get established.

- The TCP connection will be established using a three-way handshaking method, where the destination machine, Host B, will listen the request from Host A and will acknowledge it with a response.
- Once the TCP connection is established, it will be maintained until the data communication is complete between Host A and B.
- Application data as per the user's application on the computer will get converted into logical segments by the transport layer with sequencing of the individual segments that will help during end-to-end transport.
- Segments from the transport layer to be served by the Internet layer where the encapsulation of the segment into the IP packet will take place. Every packet will have its destination IP address in the header.
- Based on the destination IP address, the routers connected in the network can take routing decision to guide the IP packet with the best network route to reach the destination network.
- The IP packet again has to undergo an encapsulation, where the IP packet will be attached with the hardware or MAC address of the source and destination, when it comes at the link layer. The data format at the link layer is called as frames.
- The frames have to get transmitted in the form of signals or bits within the network when they arrive at the physical layer.

2.5 OSI

The OSI model was developed after the TCP/IP model and is an open standard for all communication systems. The OSI model has been developed by the International Standard Organization [1,8]. However, it is very interesting to know that the protocols associated with the OSI model are not used commonly, the OSI model itself is quite general and valid, and the features which are described at each of its layers are still very important. The TCP/IP model on the other hand has the opposite properties: the model itself is not commonly used, but TCP/IP protocols are widely used.

2.5.1 Layered Architecture

The OSI model has seven layers, which we will discuss below. These layers are further divided into two groups which are "application support layer" and "network support layer". The upper three layers which comprise the session, presentation, and application layers of the OSI reference model have processes which support user's application and hence have been grouped as the application support layer. And the lower four layers comprising the physical layer, data link layer, network layer, and transport layer are the network support layers [6, 9–11].

The seven layers of the OSI model as represented in Figure 2.6 are described below:

- Application
- Presentation
- Session

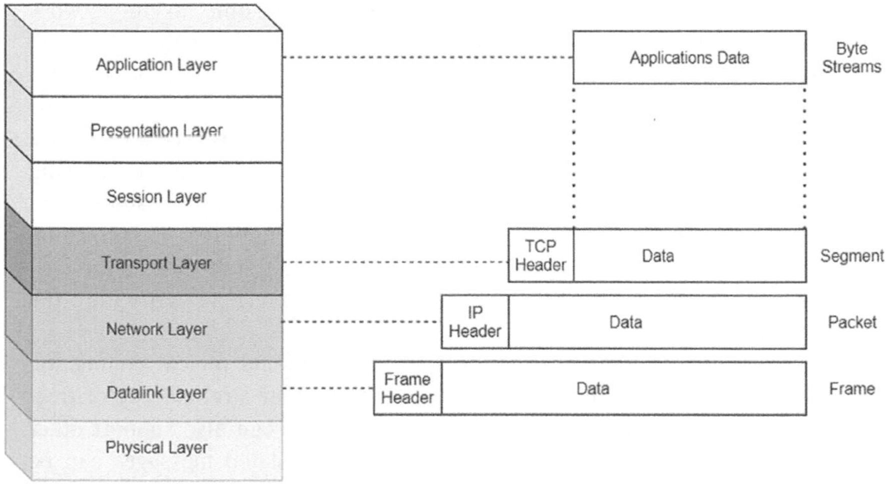

FIGURE 2.6 OSI model layered architecture.

- Transport
- Network
- Data link
- Physical

Application Layer: The application layer is responsible for providing of the interface to the user to communicate through the computer as an access to the network. This layer has to encompass those protocols which directly interact with the user, like the HTTP that is needed for the World Wide Web, FTP, electronic mail, and so on.

Presentation Layer: The presentation layer is responsible for the presentation of the data as per user's application and by that way it provides service to the application layer. The important duty of the presentation layer is of data translation, and code formatting to define the data in the native format of the remote host should be presented in the native format of the host.

Session Layer: The session layer is responsible for maintaining of sessions between remote hosts. The duty of the session layer is of setting up, managing, and then tearing down sessions between two of the interacting machines.

Sessions provide various services, including dialog control of the devices (keeping track of whose turn it is to transmit), token management (that is used to prevent two parties from attempting of the same critical operation simultaneously), and synchronization. With that we can think of the session layer as the layer which basically keeps different application's data separate.

Transport Layer: The transport layer is responsible for successful end-to-end delivery of data from higher layers between different hosts. The duty of the transport layer is to accept data from the layer above it, split the data into smaller units, pass the transport layer protocol data units to the network layer, and ensure that all the data units arrive correctly at the other end. We can call the transport layer as a true end-to-end layer because it carries data all the way from the source host to the destination host. In the simplest way, it can be assumed as an application on the source host carries on a conversation with a similar application on the destination host, using the message headers and control messages provided by the transport layer. Considering the lower layers, each protocol is to support communication between a machine and its immediate neighbors, and not between the ultimate source and destination machines, that may be separated within the network by different routers.

The transport layer also establishes an end-to-end logical connection when it supports the higher layer or session layer for a reliable, error-free, connection-oriented transport service. However, it can also support other kinds of possible transport service, where the isolated messages can be transported with no guarantee about the order of delivery, and messages broadcasting to multiple destinations.

Network Layer: The network layer is responsible for address assignment and uniquely addressing hosts in a network and controlling operations of a network/subnets. The duty of the network layer is to provide service to the upper layers with independence from the data transmission and switching technologies which are used to connect systems; it is responsible for establishing, maintaining, and terminating network connections. It is the layer which defines how the data packets will be routed within the network, which path will be the best path to reach the destination host among the available paths. Routes can be statically defined or configured in a router's routing table or can automatically take a dynamic route with a dynamic routing configuration.

The presence of too many packets in the subnet can create congestion in the network; handling of network congestion is the duty of the network layer so as the higher layers can adapt in and will place the data load on the network accordingly. More generally, it can be said that the quality of service provided (delay, transit time, jitter, and so on) is also a network layer design issue.

Data Link Layer: The data link layer is responsible for receiving the IP packets from the network layer and translating the network layer information into data link frames. It will use MAC/hardware addressing to deliver the data link frames to the proper device on the link. This layer has the duty of providing physical transmission of the data from the network layer and handling link errors detected at this layer, network topology, and flow control. The data link frame includes a customized header with the hardware destination and source address.

The data link layer has two sublayers:

- **MAC sublayer (IEEE 802.3):** The MAC sublayer defines about contention of the media, i.e., how the packets are placed on the media. The contention of the media access is based on a "first come/first served" approach–based media access that allows everyone to share the same bandwidth. Physical addressing and the logical topologies are defined at this sublayer. Where the logical topology represents the signal path through the physical topology.

- **Logical link control sublayer (IEEE 802.2):** The logical link control sublayer defines rules to provide encapsulation as per the network layer protocol information present in the IP packet. The logical link control header has the information to guide a host with all the network layer protocol information.

Physical Layer: The physical layer is the lowest layer of the model and is responsible for transmitting raw bits over a communication channel as per the frames received from the data link layer. The physical layer deals with hardware, cabling, wiring, and power to send and receive signals in the form of bits. This layer provides specification to activate, maintain, and deactivate a physical link between the end systems.

2.5.2 OSI Operation

Consider Figure 2.7 to understand the operation of the OSI model. Each of its layers will have protocols, some of which are mentioned in the diagram, but it is not limited to these protocols only (Figure 2.7).

- The OSI model specifies that a layer on Host A speaks the same language at the same layer on Host B or with any other host in the network.

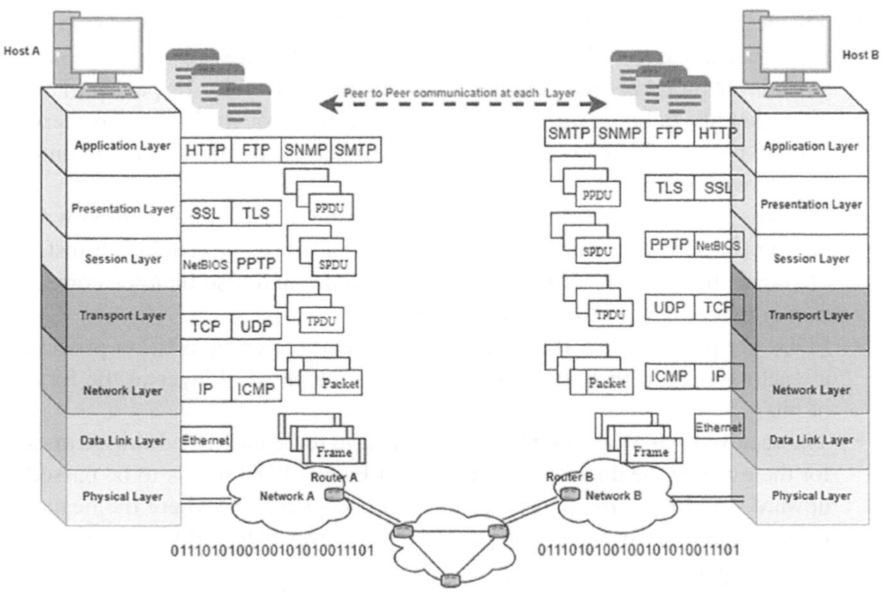

FIGURE 2.7 Operation of OSI model.

- The application layer "closest to the end user" in the OSI model receives information directly from the user's application. It uses communication through lower layers in order to establish connections with applications at the other end.
- Each layer in the OSI model adds its own information as the header to the data received from the higher layer. This header contains information specific to the protocol in operation at that layer, and this process of adding header is called as encapsulation.
- Encapsulated data is transmitted to the peering layer as protocol data units (PDUs), since the OSI model facilitates peer-to-peer communication.
- Specific to each of the layer, there are different PDUs like presentation PDUs, session PDUs (SPDUs), transport PDUs, and so on. Considering that, higher layer PDUs are encapsulated inside the PDU of the layer below that.
- The presentation layer performs the presentation or translation of application format data to network format data, or vice versa. Consider an example of data encryption and decryption that happens at this layer.
- The session layer is used to create sessions for an application to communicate with two devices in the network. It will receive a presentation PDU from the presentation layer and encapsulate it into the SPDU and keeps application-specific SPDUs separate from each other and transfer the SPDU with using the lower layers to the peer session layer on the destination host.
- The transport layer will receive the SPDU from the session layer and provide the end-to-end data transfer between end systems and hosts by encapsulating the SPDU into the transport PDU.
- The network layer will receive the transport PDU from the transport layer and encapsulate it with source and destination IP addresses and with network layer protocols information in its header to form an IP packet. This IP packet has to be forwarded, including routing through different routers. This IP packet will be processed at the destination network router, and until then, it will be forwarded.
- The data link layer performs encapsulation of the IP packet into frames with adding the hardware source and destination addresses. It provides transfer of frames between two directly connected nodes and also handles correction of the error from the physical layer.
- PDUs are passed down through the layers with the encapsulation process at each layer until they are transmitted over the physical layer in the form of bits.
- The received bits from the physical layer have to be converted into frames for their use by the data link layer on Host B. The PDU that is to be passed upward is to follow the reverse process of encapsulation, where the header information is to be stripped to get the higher layer PDU before passing the layer above it.
- This process has to continue until all the PDUs are successfully received and error-free at the destination host application layer.

2.5.3 Comparing TCP/IP and OSI Model

Even though there are so much in common between the OSI and the TCP/IP model, there are certain differences as well. We are here comparing both models in below-mentioned points [12,13] (Figure 2.8).

- OSI is considered as a more generalized, protocol-independent standard, where the model was not biased toward a specific set of protocol. However, the TCP/IP model is considered as the model which is based on standard protocols that had helped the Internet to develop.
- The OSI model has the protocols hidden when compared with the TCP/IP model, and therefore, they can be replaced relatively easily as per the change in technology. One of the main purposes of having layered protocols in the first place is to cope with those technological changes transparently.
- OSI has designed seven layers in the model when compared with TCP/IP model; there is no concept of a separate presentation and session layer.
- Being the most experienced protocol sets in the world, the TCP/IP is still more widely accepted in the communication industry than the TCP/IP model itself. On the other hand, the OSI model has all the definitions of the layered service to make it the recommended industry standard model.
- The TCP/IP model has protocol-centric approach, which means the protocols came first and then the model was defined as per the protocol, but it suffers to accommodate new protocols. However, the OSI model has model definitions first, and the inclusion of protocols was done later and was having problem in fitting a standard protocol into the model at the time of network implementation of the OSI model as the networks were not matching with desired specification.
- Comparing on the grounds of connectionless versus connection-oriented communication, we can find that the OSI model supports both connectionless and connection-oriented communication in the network layer, and it supports only connection-oriented communication in the transport layer.

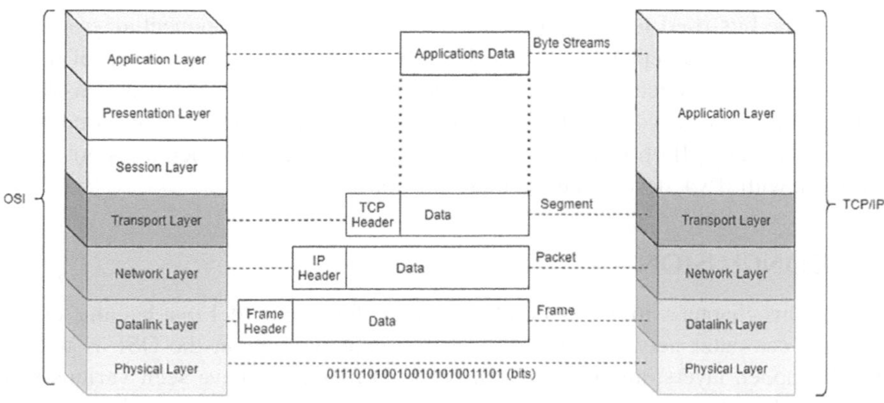

FIGURE 2.8 Comparison of OSI and TCP/IP.

However, the TCP/IP model supports only one connectionless mode of communication in the network layer but supports both in the transport layer, giving the users a choice of using any of the mode of communication.

2.6 IP ADDRESS

The IP address is the addressing mechanism used at the network layer, and every host, computer, router, or any device on the Internet has an IP address. This address can be used in the source address and the destination address fields of IP packets and works as a numeric identifier of a machine. This addressing is logical addressing that means it is a software address not a hardware address. It is a numerical label assigned to each device connected to a computer network using the IP for communication. All the devices which are connected to the Internet should have a unique IP address that will be used to identify the device on the Internet, which means that there is always a need of billions of IP addresses and will be increasing as per the increase of devices/gadgets connecting Internet.

2.6.1 IPv4 Address

IP Version 4 (IPv4) has 32 bits of the number to be used as an IP address for any host or in a network. These 32 bits can be represented in a dotted decimal format of numeric representation. These 32 bits are further divided into four octets or bytes separated by a dot, each containing 8 bits and each of the 4 bytes is written further in decimal, from 0 to 255.

The IP addresses are hierarchical with each IP address comprising a variable-length network portion in the top bits and host portion in the bottom bits. On a single network, the network portion will have same bits for all hosts in that network, e.g., Ethernet local area network.

2.6.2 IPv6 Address

With the depleting number of the IPv4 address, it was observed as a need to have billions of IP addresses to be available globally. That can be possible only if we can increase the bits used for IP addressing. Hence, due to the exponential growth of the Internet and the depletion of available IPv4 addresses, a new version of the IP (IPv6) was standardized in 1998 using 128 bits for the IP address. It is considered as a replacement of IPv4, and the Internet had allowed IPv6 addressing for communication. However, IPv6 is a different network layer protocol which does not really interwork with IPv4, despite having many similarities.

2.7 CONCLUSION

Starting this chapter with the need of standard reference model that became essential for a computer network. We have further explained about the OSI as well as TCP/IP model, layers, and comparison between them. We have seen various protocols and how important they are in a computer network to accomplish successful communication.

REFERENCES

1. Forouzan, B. A., (2021). *Data Communication and Networking*, New York: McGraw Hill.
2. Stallings, W., (2006). *Computer Organization and Architecture*, Hoboken, NJ: Pearson.
3. Wang, S. P. (2021). Communication, TCP/IP, and Internet. *Computer Architecture and Organization* (pp. 243–292). Singapore: Springer. [online]. Available: https://www.springerprofessional.de/en/communication-tcp-ip-and-internet/19912034
4. Priya, B., What is client/server network and types of servers?, Sep. 15, 2021. tutorialspoint. Accessed: Oct. 5, 2021. [online]. Available: https://www.tutorialspoint.com/what-is-client-server-network-and-types-of-servers
5. Hunt, C., (2002). *TCP/IP Network Administration*, Sebastopol, CA: O'Reilly Media, Inc.
6. Stevens, W. R., (1994). *TCP/IP Illustrated, Vol. 1: The Protocols*, Boston, MA: Addison-Wesley Professional.
7. TCP/IP protocols, IBM Corporation. Accessed: Mar. 05, 2021. [online]. Available: https://www.ibm.com/docs/en/aix/7.2?topic=protocol-tcpip-protocols
8. Tanenbaum, A. S., Feamster, N., Wetherall, D. J. (2021). *Computer Networks*, New York: Pearson.
9. TCP/IP Protocol Architecture Model, Oracle Corporation. Accessed: Feb. 24, 2021. [online]. Available: https://docs.oracle.com/cd/E19683-01/806-4075/ipov-10/index.html
10. Tucker, C., The OSI Model – The 7 Layers of Networking Explained in Plain English, Dec. 21, 2020. freeCodeCamp. Accessed: Feb. 21, 2021. [online]. Available: https://www.freecodecamp.org/news/osi-model-networking-layers-explained-in-plain-english/
11. Layers of OSI Model, Jan. 12, 2022. GeeksforGeeks. Accessed: Feb. 1, 2022. [online]. Available: https://www.geeksforgeeks.org/layers-of-osi-model
12. OSI vs TCP/IP, JavaTpoint. Accessed: Mar. 5, 2021. [online]. Available: https://www.javatpoint.com/osi-vs-tcp-ip
13. Ginni, Difference between OSI and TCP/IP Reference Model, May. 4, 2021. tutorialspoint. Accessed: May. 28, 2021. [online]. Available: https://www.tutorialspoint.com/difference-between-osi-and-tcp-ip-reference-model

3 The First Internet *ARPANET*

ABBREVIATIONS

ARPANET	Advanced Research Projects Agency Network
BBN	Bolt, Beranek and Newman
CMU	Carnegie Mellon University
DARPA	Defense Advanced Research Projects Agency
DDN	Defense Data Network
ICCC	International Conference on Computer Communications
IMP	Interface Message Processors
IPTO	Information Processing Techniques Office
MIT	Massachusetts Institute of Technology
NCP	Network Control Protocol
NCSA	National Center for Supercomputing Applications
NSF	National Science Foundation
NWG	Network Working Group
PAD	Packet Assembler/Disassembler
RFC	Request For Comment
RP	Responsible Person
SDC	System Development Corporation
SRI	Stanford Research Institute
TAC	Terminal Access Controller
TIP	Terminal Interface Processor
UCLA	University of California at Los Angeles
UCSB	University of California of Santa Barbara
WWW	World Wide Web

3.1 INTRODUCTION

"Internet" is the most common and popular technology today that everyone must be familiar with, considering it has gained so much importance to us. The evolution of this giant worldwide network that we call "Internet" has not been a single-day story, and not a single person can be said as the father of Internet. The reason is that it holds a long list of pioneers whose contributions are important to its development. This includes a number of scientists, researchers, and scholars who were part of the development. From ideation to formulation into concepts then to implementation to operationalization, its journey has been very interesting historically. The development of ARPANET network was a milestone in its journey and is considered the birth of Internet [1–9].

DOI: 10.1201/9781003302902-3

3.2 ARPANET

The Advanced Research Projects Agency Network (ARPANET) has been the first network based on wide-area packet switching network with distributed control, established by Defense Advanced Research Projects Agency (DARPA) of the United States Department of Defense. It has been also considered one of the first networks to implement the TCP/IP protocol suite. Both of these technologies, that is, the packet switching network and the TCP/IP protocol suite later became the technical foundation of the Internet. This network was originally setup as an experimental laboratory and was used for Department of Defense computer science and networking research. It was declared as an unclassified, packet-switched data network later by Defense Communication Agency. The ARPANET is now one of the subnetworks of the Defense Data Network (DDN) and is managed by the Defense Data Network Program Management Office (DDN PMO). DARPA established policy for the ARPANET and also decided who to become subscriber of ARPANET [10]. As such, the subscribers were required to follow certain technical and administrative procedures to connect their host computers or other equipment to the DDN. The procedures document was given a public release as an unclassified document describing background and technical information on the ARPANET.

3.2.1 A Brief History of ARPANET

Dr. Joseph Carl Robnett Licklider, also famous as J. C. R. or "Licklider" working at BBN Cambridge, was appointed as the head of the Information Processing Techniques Office (IPTO), an office sponsored by the U.S. Department of Defense Advanced Research Projects Agency (ARPA). Licklider became the first director of IPTO in 1962 with taking 2 years leave from BBN in Cambridge, Massachusetts to give guidance to this newly created office, where he guided ARPA's funding of computer science research. In 1962, he formulated the earliest ideas of global networking that he wrote in a series of memos with the discussion on "Intergalactic Computer Network". This work is considered the first recorded description of the social interactions that could be enabled through networking and the concept presented was very much similar to the Internet of today. His idea was envisioned for having an interconnected set of computers with globally located sites through which everyone could quickly access data and programs from any of its sites. Licklider was trying to define the problems and benefits resulting from computer networking, while at IPTO, Licklider was able to convince his successors at ARPA, Ivan Sutherland, Bob Taylor, and MIT's Lincoln Labs researcher Dr. Lawrence (Larry) G. Roberts, about the importance of this networking concept. Licklider formed his group of individuals as well as a number of forward-thinking corporations to bring this grand new world of computer networking into existence [11].

Licklider left IPTO for IBM after 2 years in July 1964 with Dr. Ivan Sutherland, a young prescient computer scientist and a member of Licklider's network, becoming the new director of IPTO. One of the first decisions of Sutherland after taking over as the director of IPTO was to fund the work of his friend Larry Roberts, who had known each other since entering their MIT doctoral program in 1959. Roberts had attended MIT, from where he received his bachelor's degree (1959), master's degree (1960), and PhD (1963), all in electrical engineering. Roberts further continued to work at the MIT's Lincoln Laboratory in the field of computer graphics after receiving his PhD in 1963. Roberts also started working in the field of computer-to-computer packets data networks, having read Licklider's 1961 paper "Intergalactic Computer Network".

Licklider, Sutherland, and Roberts were part of the ARPA scientist's contingent traveling by train to Hot Springs, Virginia, to attend the Second Congress of the Information System Sciences sponsored by MITRE Corporation, in November 1964. In the history of computer networks, it was the one such event where it was discussed that the most important problem in the computer field was related to computer networking; the ability to access one computer from another easily and economically to permit resource sharing. One year later, Donald Davies from National Physical Laboratory, UK documented a concept of a store-and-forward system for very short messages (which is now called as packet switching) presented as the ideal communication system for interactive systems. In June 1966, Davies wrote a second internal paper in which he coined the word packet, mentioning it as a small subpart of the message the user wants to send. Davies also introduced in the same paper about the concept of an "interface computer" to sit between the user equipment and the packet network.

In 1966, Robert Taylor took over as the director of ARPA's Information Processing Techniques Office (IPTO). He noticed that most of contractors at IPTO research were constantly requesting for more computing resources and wanted their own computers, which was an expensive luxury. He also noticed that there was a lot of duplication of research work getting done at IPTO and was wasteful of resources and money. Taylor decided that ARPA should link the existing computers at ARPA-funded research institutions together, considering the backgrounds from theoretical legacy of Licklider. The purpose was to allow everybody on the network to share computing resources and results. With this idea to go-ahead with building of such a network, Taylor started looking for someone to manage this project. With his search on the project, Taylor's first choice was namely the young Larry Roberts, who was well respected in his field and known for his good management skills and dedication to his work.

Prior to this in 1965, Tom Marill, a student of Licklider been influenced by his interest in computers, approached ARPA. Tom Marill expressed his desire of conducting an experiment linking (MIT's) Lincoln Lab's TX-2 computer to the SDC Q-32 computer in Santa Monica. The ARPA officials perceived it as a good idea but suggested Marill to carry out his experiment under the sponsorship of the Lincoln Laboratory. This experiment happened under sponsorship of Lincoln Lab with Larry Roberts as the in charge of the project. This experiment was although much smaller than the ARPANET in scope but was a success. However, the system response times were slow and connection reliability was often poor, but this project had provided a solid first step in this direction. Based on that, Roberts and Marill published a paper in 1966 about their success at connecting over dial-up [12].

Taylor hired Larry Roberts (who later became the director of ARPA, IPTO) as program manager and technical architect for the ARPANET, the precursor to the Internet in 1966. APRANET was designed by Roberts [6], and he managed its building for over next 6 years. Roberts attended a meeting for ARPA's Principal Investigators (PIs) – the scientists heading ARPA-sponsored research projects. Where he laid out his plans to connect all of the ARPA-sponsored computers directly over dial-up telephone lines. Roberts earlier considered networking all the host computers with the host computers to handle the network functions at each site. This idea was not well received, and he had to adopt a concept of interface message processors (IMPs), a small computer to handle network functions. It was further decided by Roberts to have start with 4 ARPANET network sites initially at UCLA, the Stanford Research Institute (SRI), the University of Utah, and UC Santa Barbara. Considering this as the core of the network and the network could grow from there. The bid was requested for building of the IMPs to more than 140 companies and Bolt, Beranek and Newman (BBN) was selected for building of the IMP by December 1968. BBN delivered the first IMP to the ARPANET site at UCLA in August 1969 and a month later the second IMP at SRI site. With connecting both of the IMPs at these sites, it was considered the birth of ARPANET. The journey of ARPANET started from there as an experimental packet-switched host-to-host network in late 1969 as shown in Figure 3.1 the ARPANET with its four sites connected with IMP located at each of its sites. ARPANET provided successfully efficient communications between different host computers at its multiple sites, allowing them to have convenient sharing of hardware, software, and data resources among a varied community of geographically-dispersed users. By April 1971, the ARPANET had grown to 15 sites and 23 host computers (Figure 3.2).

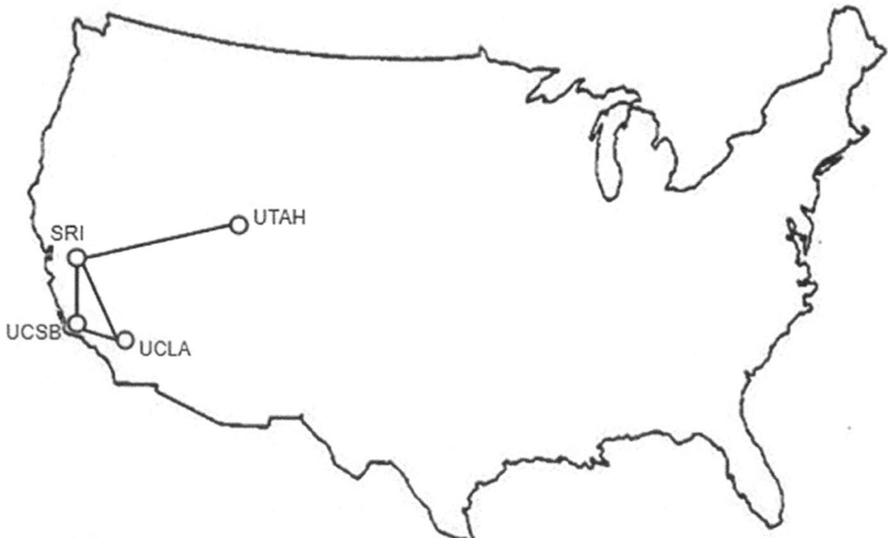

FIGURE 3.1 ARPANET sites connected with IMP in 1969.

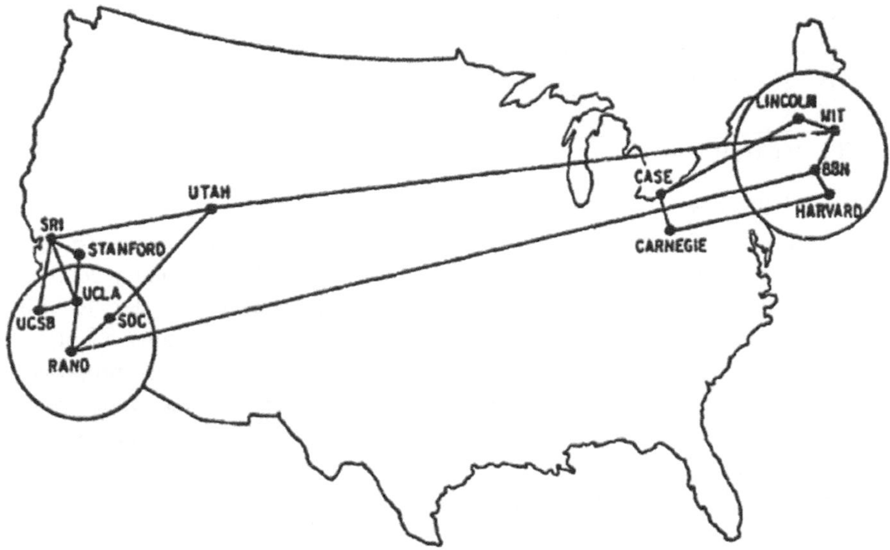

FIGURE 3.2 Sites connected with IMP in 1970 (ARPANET 1970).

3.2.2 Sites of ARPANET

With the great success of ARPANET's operation, Lawrence Roberts met with the Network Working Group (NWG) in Utah in 1969. He emphasized the need for further achievement in the direction of developing a viable network protocol. This group was able to create a Network Control Program (NCP) by 1971, and that became the standard networking program for the ARPANET. NCP provided several standard network services, the set of protocols that could allow sharing of network services by several applications running on a single host computer. The purpose of NCP was to provide the feature of establishing, breaking, and switching of connections, and controlling of the flow of communication between different host computers.

The following were the initial 15 sites of ARPANET (by April 1971) connected to the NCP:

1. University of California at Los Angeles (UCLA)
2. Stanford Research Institute (SRI)
3. University of California of Santa Barbara (UCSB)
4. University of Utah
5. Bolt Beranek and Newman (BBN)
6. Massachusetts Institute of Technology (MIT)
7. RAND Corporation
8. System Development Corporation (SDC)
9. Harvard University
10. Lincoln Laboratories
11. Stanford University
12. University of Illinois, Urbana

13. Case Western Reserve University
14. Carnegie Mellon University (CMU)
15. NASA-AMES

3.2.3 Motivations

3.2.3.1 Time-Shared Computers

It was observed as the need of sharing the computer resources among different users at locations different from the host computers. For that, the development needed was to have the interconnection available among those independent computers to fulfill this requirement. Roberts presented this requirement into one of his paper "Computer Network Development to Achieve Resource Sharing" in 1970 [13]. He proposed a computer network for a set of autonomous and independent computer systems, and they were supposed to be interconnected so as to allow interactive resource sharing between any pair of systems. He mentioned that the basic technology required to create such a network to share computer resource was available with the advent of time-sharing. Users through remote consoles can share all the resources available to a time-sharing system. This was based on the concept of splicing the two systems together as remote users of each other and then the user programs were allowed to interact with them. An experiment was carried out based on that characteristic between the Lincoln Lab's TX-2 computer and the SDC's Q-32 computer in 1966.

What was observed as obstacle to the supposed to be resource-sharing network after the analysis of that experiment, and why it became a requirement to search for a new communication network was further mentioned by Robert. Some of them were as mentioned below:

- The console-based interconnection was logically a good approach to share the resources of one system to another, but practically interconnection with console grade communication service was virtually useless.
- Worth of network can only be determined by the maximum number of worker systems supported on the network, as if the network supporting only two systems will be worthless compared to the network supporting twenty systems.
- The response of telegraph or voice grade communication lines for network connections had certain service degradation and were not encouraging to the network users.
- It would be costly to have nationwide interconnection of computers based on leased lines or dial-up lines.

3.2.3.2 Network Working Group

Researchers and experts from four of the IPTO funded research laboratories were called together by ARPA in 1968. The agenda was to identify and solve the technical problems associated with the development of the ARPANET. The meeting was organized by Elmer Shapiro from the Stanford Research Institute (SRI) as per instructions from ARPA, with discussions held on host-to-host problems. Steve Crocker represented UCLA, Steve Carr represented University of Utah, Jeff

Rulifson represented SRI and Ron Stoughton represented UCSB. As the outcome of this meeting, it was decided that the meeting should happen regularly to discussion on the issues and obstacles of ARPANET, and this group was known as the Network Working Group. The Network Working Group (NWG) started meeting regularly to discuss the progress of their work, technical standards, network design and architecture, and many other aspects of computing and networking. NWG provided a common platform for interaction among the research groups from various organizations and the ARPANET research work was in full progress.

There were many problems the NWG started discussing like the programming and other technical problems to enable different computers to communicate with each other. Also, the group observed that there was a need for an agreed set of signals that would allow the communication channels open up, would then allow the exchange of data, and then would close the channel after the exchange of data completed. These agreed set of rules were called as the protocols. ARPANET required that the sites to work together for establishing necessary protocols.

3.2.3.3 Request For Comments (RFC)

With the formation of the Network Working Group, the group started meeting regularly to speed up the progress, research, and solution to the obstacles. The Network Working Group realized during a meeting in Utah in February, 1969, that they need to start writing down their discussions. Steve Crocker used the term "Request For Comments" first time when he was volunteering to organize the notes written by the Network Working Group. The first RFC entitled "Host Software" was written by Steve Crocker on April 7, 1969. Crocker along with Vinton Cerf, Jon Postel, Bill Naylor, and Mike Wingfield was responsible for creating the ARPANET protocols [14], and these protocols later became the foundation of present day's Internet.

The Request For Comment (RFC) became a very useful and convenient method for the purpose of recording all the details and technical information in the research carried out by the Network Working Group. With that, the RFC became the official document of the Network Working Group.

3.2.3.4 Network Control Protocol

The Network Control Protocol (NCP), also called as Network Control Program, refers to the software program of ARPANET transport protocol that was built for applications on ARPANET hosts. It provided the host-to-host transport service with connections and flow control between processes running on different ARPANET hosts. NCP was later replaced by TCP/IP protocol in the 1980s. The Network Working Group (NWG) under Steve Crocker finished the initial version of host-to-host protocol of ARPANET in December 1970, they called it as the Network Control Protocol (NCP). After completing the implementation of NCP on ARPANET sites during 1971–1972, the network users started working on developing applications on top of NCP. This allowed ARPANET users to access the host computers at remote sites and transmit files between computers. It provided the middle layer of the protocol stack and supported applications such as email services and file transfer services.

NCP was the host-to-host protocol and acted much like a transport layer in TCP/IP stack, as the procedures were defined for connecting two hosts on ARPANET. The

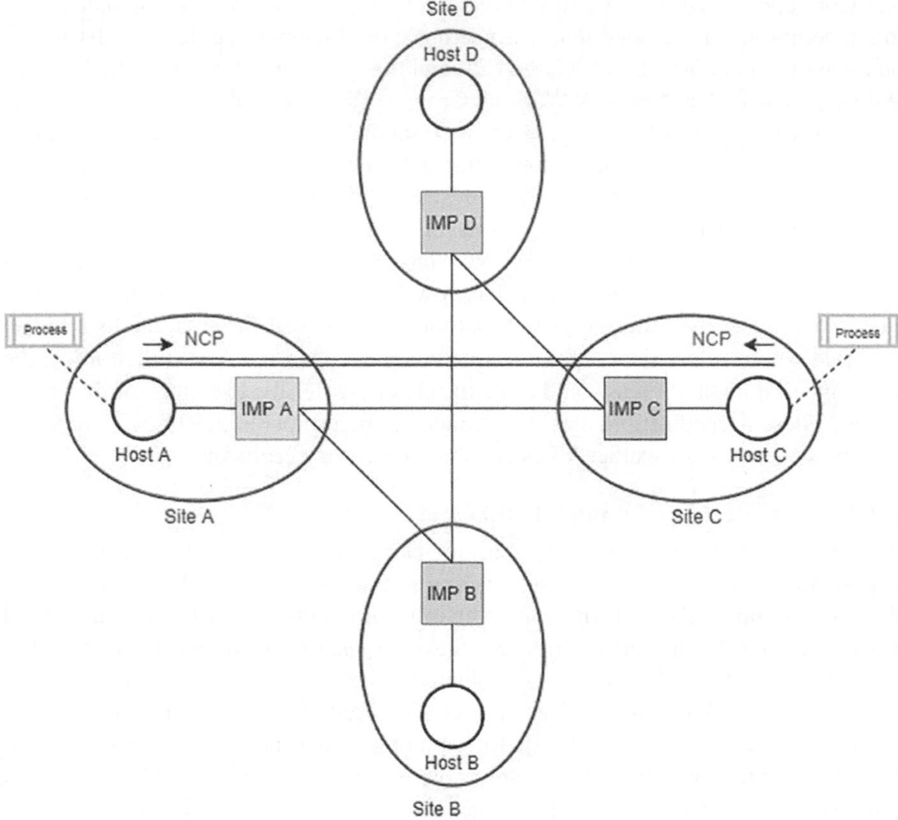

FIGURE 3.3 ARPANET sites with NCP host-to-host connection.

transition from NCP to TCP/IP is marked as a major move toward the modern Internet in 1983 and TCP/IP remains the standard protocol of the Internet communication (Figure 3.3).

3.2.3.5 Decentralization of Network or Distributed Control

Paul Baran from RAND Corporation had published a report "On Distributed Communications Networks" in 1962, and his research was funded by the U.S. Air Force [15]. Baran discussed about the protection of the communications systems of U.S. military from serious attack based on the principle of "redundancy of connectivity". He explored several models of forming communications systems and evaluating their vulnerability in that report. In that report, he proposed a communications system without a central command and control point. It was outlined that all of the surviving nodes would be able to reestablish connection in the event of an attack at any point and the whole network would remain un-impacted with damage to any part of it. This will minimize the effect on the whole network due to any attack. The focus was to increase the decentralization in the network as much as possible rather than a central control for the purpose of network's survivability. As the centralized network is vulnerable, the destruction of the central node can destroy the interconnection between end nodes.

He mentioned the need for a perfect switching in a distributed environment with flexible switching and low cost links. The network to have survivability in the event of a damage should allow the switching path to be dynamic so that switching scheme can find any possible path that might exist after a heavy damage to the network. The need to explore the possibilities of a "real-time" data transmission system based on store-and-forward techniques. The network path selection should be able to search and select the shortest path possible instantaneously available in a rapidly changing network.

3.2.3.6 Store-and-Forward Switching

The discussion started in the 1960s to adopt a new public utility switching methodology for data communication other than the traditional public utility of voice communication's circuit switching. With a strong opposition from the computer research community and industry experts, some felt the need of having something in this direction that became famous as the "packet switching". A new type of switches were proposed for the data communication networks. The purpose was to leverage the switching devices with a store-and-forward technology [16]. The term "Packet" was first used by Donald W. Davies in his 1966 paper "Proposal for a Digital Communication Network". He referred a packet as a small subpart of the message that the user want to send through the network. It was envisioned that a communications network to use trunk lines (T1) from 100 kB/sec in speed to 1.5 MB/sec and the message sizes of 128 bytes, with that a switch could handle up to 10,000 messages/sec. It was his observation that the ideal data communication network should have short messages system with store-and-forward mechanism. He also mentioned in his paper about the concept of an "interface computer" that he proposed to sit between the user equipment and the packet network.

Packet networks utilized the concept of interface message processors (IMPs) in the network design to provide packet switching capability. The IMPs were connected with leased telecommunication lines to provide store-and-forward capability to very short messages. In the ARPANET, these IMPs were minicomputers at every node served by the network, which were interconnected in a fully distributed fashion by 50 KB leased lines.

The ARPANET has each of its minicomputer taking blocks of data from the computers or terminals connected to it. It further subdivided the data block into 128-byte packets, with adding a header specifying destination and source addresses to the data packet. The minicomputer was dynamically updating the routing table based on that minicomputer sent the packet over whichever free line was currently the fastest route toward the destination. The next hop minicomputer have to acknowledge it and to repeat the routing process independently upon receiving a packet. This was one of the important characteristics of the ARPANET, and that was based on completely distributed, and dynamic routing algorithm on a packet-by-packet basis. That dynamic routing algorithm was based on a continuous evaluation within the network with the least delay paths, considering both line availability and queue lengths.

3.2.3.7 Responsiveness of Network

It was Larry Roberts' analysis regarding the responsiveness of the network that the total round-trip delay of the network must not exceed the human short-term memory span of one to two seconds where a user is making more or less direct use of a complete remote software system. The time-sharing systems were probably introducing delay of at least one-second, and so the network's end-to-end delay should be less than half a second. This was his analysis that the network response time should also be comparable, if possible while using a remote display console over a private voice grade line. The network should be able to send interactive graphics if available with a complete new display page requiring about 20 Kb of information within a second. The network should have shorter message delay where two programs are interacting directly without a human user getting involved and the job will get through sooner. A user will have to duplicate the remote process or data at his site if the communications system substantially slows up the job. A reasonable measure of network responsiveness, to compare communications systems performance was proposed as the "effective bandwidth" by computing "data block length for the job/end-to-end transmission delay". The success of the ARPANET both technically and operationally demonstrated to the world that packet switching could be organized to provide an efficient and highly responsive interactive data communications.

3.2.4 ARPANET Topology

The initial ARPANET topology is shown in Figure 3.4.

The ARPANET network topology was designed by Roberts and presented in his paper "Computer Network Development to Achieve Resource Sharing" in 1970. Topology of the network was considered with the purpose of minimizing cost of

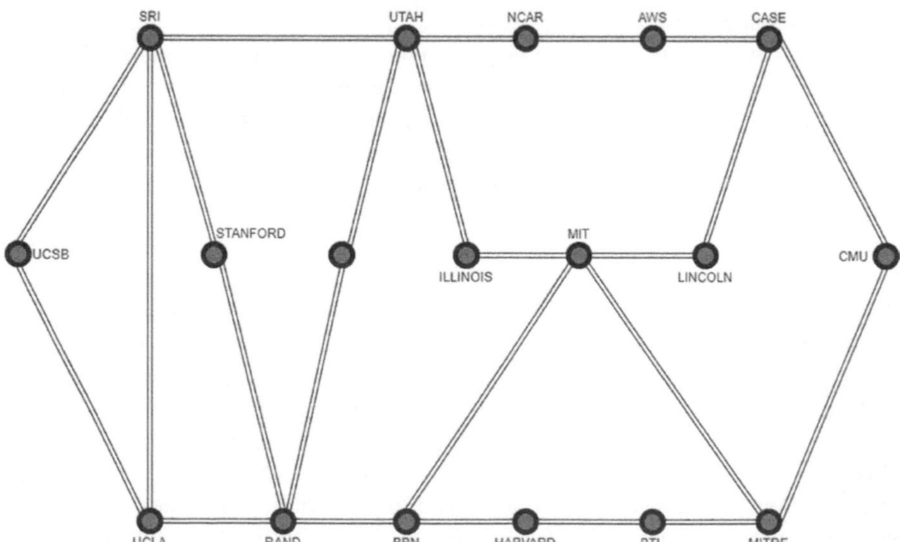

FIGURE 3.4 ARPANET initial topology diagram.

the transmission lines, maximizing of the growth potential, and to satisfy the design criteria. Roberts consulted Dr. Howard Frank, who had founded Network Analysis Corporation (NAC), a company specialized in network topology design. Capacity analysis of the ARPANET was performed by Network NAC during the optimization of the network topology. As per the analysis, the network was having the capability of flexibly increasing its capacity by the addition of additional transmission lines. Cost-performance of the network can be improved with using of 108 and 230.4 Kb communication services, wherever appropriate.

3.2.5 ARPANET Network Operation

ARPANET network was designed with the mechanism of providing store-and-forward communication paths between host computers distributed across the continental US. The task of message handling was performed by a special purpose computers connected to the host computers at each site to handle networking functions. These small computers were called as the Interface Message Processor (IMP) in the ARPANET.

The host computers at each site were connected through the IMPs with a full duplex telephone lines of 50 kilobits/sec capacity as shown in Figure 3.5. All of these

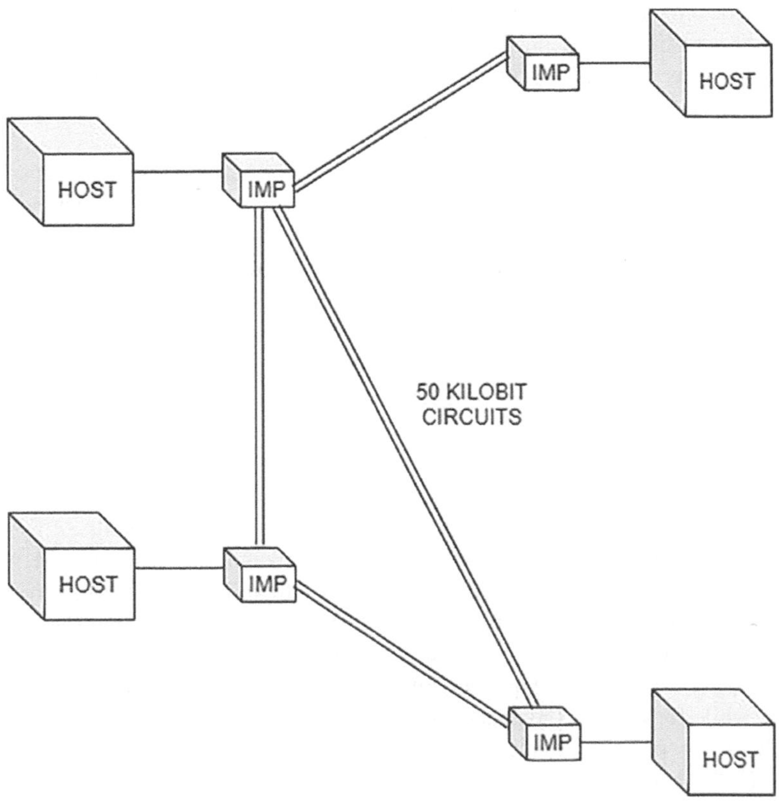

FIGURE 3.5 ARPANET: host-IMP connectivity.

IMPs were able to speak the same language to facilitate the communication between them. Each of the host computers had to adapt only the language once in communicating with the connected IMP. This IMP was working as a gateway to the host computers, providing network connectivity to the host computers. These IMPs were also under more direct control of ARPA as compared with the large host computers.

The ARPANET started its network operations in four sites: UCLA, the Stanford Research Institute (SRI), the University of Utah, and UC Santa Barbara. This was considered the core of the network, and the ARPANET could grow from there.

A message from the host computer which was ready for transmission through the network was broken into a set of packets. Each of the packets was having the appropriate header information and can make its independent way through the ARPANET nodes to its destination. A packet when transmitted between any pair of IMPs, the transmitted IMP must receive a positive acknowledgment from the receiving IMP within a predefined interval of time. If the acknowledgment is not received then the packet will be retransmitted. The system was designed to achieve a response time of less than 0.2 seconds for short messages with the goal of network design to achieve this response time with least possible cost per bit.

ARPANET network design had considered network path level redundancy to reach any of the node and also to support in the event of a failure of nodes within the network. The network design was considered with the condition that at least two of the nodes or links must fail before all communication paths between any pair of nodes to disrupt. The routing procedure was based on the assumption that for each packet the most desirable route to be selected should be the path that contains the least number of intermediate nodes from origin to destination node.

IMPs were working as the buffer in the network, where it helped for reassembly of the packet destined to the connected host and also to store and forward the packet in the network. Initially the IMPs were having 84 buffers and each of the buffer can store single packet, out of them two/third of the buffer was used for reassembly. Buffers not used for reassembly was made available for store and forward of packet. Any arriving packet with no buffer available at the receiving IMP for reassembly traffic was to be discarded with no acknowledgment to the transmitting IMP. This packet was to be transmitted again later. If no buffer available to store-and-forward traffic, then all the incoming links to become inactive.

Host computers communicate with each other via a sequence of messages, and the IMPs take the message from the connected host in segments and then forms packets from these segments with a maximum size of approximately 1,000 bits.

The IMPs then send the packet into the network, where the receiving IMP had to reassemble the packets received and then to deliver them in sequence to the receiving host computer. The message handling was done in such a manner that the transmitting host attached an identifying leader to the beginning of each message. The IMP then attached a header to each packet with adding more information to be used in the network while forming of the packet from the message (Figure 3.6).

Each of the packets was to be routed individually within the network from IMP-to-IMP toward the destination. At each of the IMPs, the error detection mechanism was working with parity checks, initial and terminal framing characters to be shipped along with the packet for the next hop IMP. If the framing was disturbed

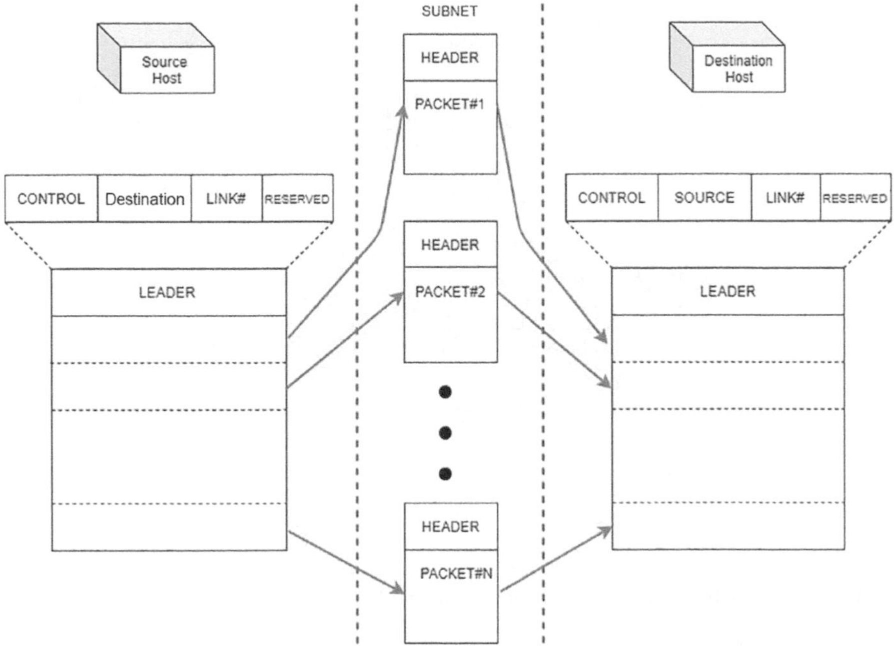

FIGURE 3.6 ARPANET: message to packet.

due to error, then the packet either will not get recognized or will be rejected by the receiving IMP.

Routing algorithm used in the network to determine the best route with smallest total estimated transit time for the packet toward the destination. When a packet arrive at an IMP, the IMP had to decide which of its output lines to transmit a packet addressed to another destination. Route selection is done based on a fast and simple table lookup procedure, and for each of the possible destinations, an entry is made in the routing table to determine next hop. The routing table keeps updated dynamically and consistently on timely basis for the changing conditions in the network.

The packets as were moving from different subnets, and each of the IMPs had to store the packet until a positive acknowledgment was received from the succeeding IMP. This indicates the message got received without any error and accepted by the receiving IMP. Similarly, the receiving IMP will store the packet until it will receive a positive acknowledgment from the next receiving IMP. If acknowledgment not received within a set time interval, then the packet has to be transmitted again by the transmitter IMP probably through different route. Packets usually arrive out of order at the destination IMP, which are further to be reassembled and header to be stripped off each packet, and a leader identifying the source host and the link that is followed by the message is then send to the destination host.

In the later development of ARPANET, a Terminal Access Controller (TAC) device was also attached to the network, and the access network with TAC is shown in Figure 3.7. TAC was a computer system attached directly to the network that lets a user at a terminal access the network without first going through a local host. TAC,

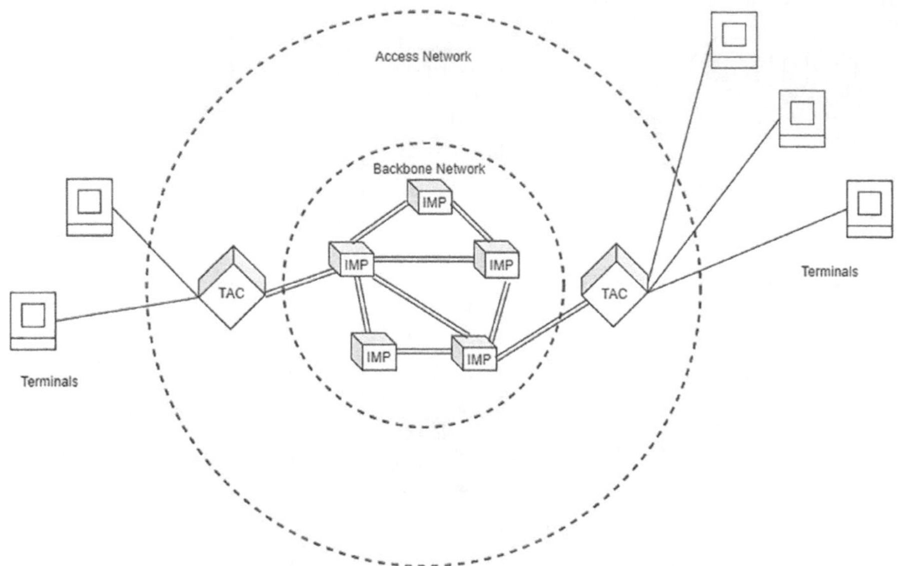

FIGURE 3.7 ARPANET: TAC-based access network.

a special terminal access node, let a terminal connect directly to the network without going through another host. The users had to be authorized for network TAC access by a DARPA-appointed network contact known as a "Responsible Person" (RP) a person in a position of authority within each organization authorized to use the ARPANET.

3.3 RECEPTION, SUCCESS, AND ACHIEVEMENT

In October 1972, the first public demonstration of ARPANET happened at the first International Conference on Computer Communications (ICCC) in Washington, D.C. The demonstration with complete ARPANET node installation was organized by Robert Kahn of BBN. Robert gave demonstration with 40 active terminals permitting access to dozens of computers from all over the United States. This was for the public as a demonstration to the attendees of the proof that how the packet-switched network really worked. This had shown the world that the packet switching is the mode of communication for the data computer networks. At this time, many of the attendees including the experienced computer professionals were finding it difficult in accepting the packet switching technology. They were not convinced with fact that the packet switching technology having collection of computers with wideband circuits and IMP switching nodes could all function together reliably. This public demonstration of ARPANET in the conference displayed the operations and reliability of packet-switched network lasted for 3 days.

The reception of packet switching (ARPANET) initially from the computer industry was generally negative due to its different approach from traditional circuit switched telecommunication network. Many of the communication professionals provided sharp criticism and expressed their anger on the use of buffers for store-and-forward mechanism used in the packet switching. Many had the opinion of the network infrastructure development to become very costly with using this and the packet network will never to be economical for the general purpose public data communication.

Fundamental to the ARPANET was the discovery of a new way of looking at computers, where the developers viewed the computer as a communications device rather than only as an arithmetic device. This new view made the implementation of the ARPANET possible. This view came from the research conducted in academic computer science that led a shift in understanding the role of the computer. The research and development of ARPANET had provided a rich legacy for the further advancement of computer science and had shown that how significant it is to study the experience from ARPANET to be used to further advance the study of computer science.

3.3.1 Growth of ARPANET

After the first four nodes of ARPANET were brought up and tested in December 1969, the network started growing very rapidly. By December 1970, the network had already grown up to 10 nodes with 19 host computers. By April 1971, the ARPANET had 15 nodes with 23 host computers. It became evident by this time that the connecting terminals directly to the network should require a Packet Assembler/Disassembler (PAD) type device, and in 1970–1971, such a device was designed and built for ARPANET. A variant of the IMP as the Terminal Interface Processor (TIP) was also designed, and the first TIP was added to the network in August 1971. The TIP was used to permit the users to get on the ARPANET network directly with their terminals. With that, the user could have direct access to all of the machines rather than having to go through one machine to get anywhere.

In 1972, network traffic on ARPANET had also grown rapidly from 100,000 to 1,000,000 packets/day that exceeded the past estimations made in 1967 by the computer experts. In 1973, the first satellite link was added to the ARPANET with a TIP in Hawaii, and later in the year, a pair of TIPs were further added in Europe.

In September 1973, the ARPANET had grown up to 40 nodes, 45 host computers having the internode traffic about 2,900,000 packets/day. By then, the ARPANET had clearly reached a stage of operational stability, heavy usage, and was by all measure can be considered a major success. In June 1974, the ARPANET had 46 IMPs, and in July 1975, it numbered with 57 IMPs. In 1981, ARPANET had 213 host computers connected, with another host connecting to it in approximately every 20 days. The ARPANET network has become a utility for both the research communities funded by ARPA and the Department of Defense as a whole.

By 1984, the ARPANET had grown to more than 100 nodes. By then, the ARPANET was separated into two parts: MILNET – an operational component to serve the operational needs of the DoD, and a research component that retained the ARPANET name.

In 1990, the ARPANET was formally decommissioned, after that partnership with telecommunication industry and computer industry assured private sector expansion and future commercialization of an expanded worldwide network, which the world knows as the Internet.

3.3.2 Impact on Computer Resource and Communication Technology

Resource sharing to optimize computer resources was one of the goals of the ARPANET [16–18]. It was observed after ARPANET that many centers/users of ARPA projects did not have a computer at all but utilized computer power from resource centers that had developed. Access to such users was provided through a TIP and used one of a collection of computers. The users were happy as they need not to run their own computer now. Services provided to them had been more reliable than if they had been limited to one machine.

Also, the number of computers required now for the users had become far less through consolidation due to the statistics of large groups and the benefit of multiple time zones from Hawaii to Europe. One group of users were able to use the Illiac IV remotely from the ARPANET for large numerical problems and would either had required their own supercomputer or probably had to commute if the ARPANET was not there. On the other hand, for those individuals who needed to use unique software to the foreign hardware and were also needed access to other computers besides access to their own computers.

After comparing and cost analyzing the computer resource need of all ARPA users assuming no network with the cost of resource availability through ARPANET, the computing cost was found to be manifold expensive. With that, it became evident that huge amount of savings were possible in that era of mainframes by computer resource pooling on demand through network.

ARPANET was the milestone, and its success led to further developments of networking technologies. In the later development computers became smaller and started to be deployed in large numbers within an organization. This resulted in several computer manufacturers developing their own technologies for local area networks (LAN).

The ARPANET gave a revolutionary change in computer and communications technology. The revolution that started with an experiment in computer networking as the ARPANET grew further into a communication revolution called as the packet switching. Today, we will find that virtually all of the world is linked by a packet-switched communications service, that any terminal can now access any computer in the world irrespective of the geographic location. And, this packet-switched data network had grown up independently from the telephone network.

3.4 CONCLUSION

Global networking was a reality with the ARPANET connectivity of University College of London, England and Royal Radar Establishment, Norway, in 1973. The term Internet was born, and the ARPANET was the first workable prototype of the Internet [5, 19–21]. The technology continued to grow, and prior to this, different computer networks did not have a standard way to communicate with each other.

Robert Kahn and Vinton Cerf developed Transmission Control Protocol and Internet Protocol (TCP/IP), a communications model standard for data transmission between multiple networks.

The ARPANET adopted TCP/IP on January 1, 1983, which is considered the official birthday of the Internet. The adoption of TCP/IP allowed different kinds of computers on different networks to "talk" to each other, and all the networks could be connected by a universal language. From there, the researchers started to assemble the "network of networks" that became famous as the modern Internet.

By then, several other agencies also started developing their own networks for their researchers to communicate and share data on the network. National Science Foundation (NSF) also provided a grant to establish the Computer Science Network for the networking services to all university computer scientists in 1981. NSF intended to provide greater access to the high-end computing resources from its supercomputer centers to be shared by scientists and engineers around the country. In 1986, NSFNET [22, 23] went online and was able to connect the supercomputer centers at the speed of 56,000 bits per second, which was the speed of a typical computer modem today. NSFNET was an intellectual leap and was the first large-scale implementation of Internet, in a complex environment with many of the independently operated networks. By 1988, the NSFNET links were upgraded to 1.5 MB/sec due to its network became congested in a very short duration.

By then, the number of Internet-connected computers also started to increase rapidly, and it grew from 2,000 in 1985 to more than 2 million in 1993. In 1991, NSFNET became the first national 45 MB/sec Internet network to handle the increasing data traffic, this was the NSFNET backbone.

In 1991, the launch of the World Wide Web by Tim Berners-Lee and colleagues at CERN [24], the European Organization for Nuclear Research, Geneva, Switzerland had also a significant impact on the history of Internet. After that, the use of Web server applications and the development of many tools for organizing, locating, and navigating through information on the Web had started gaining popularity. In 1993, Mosaic was the first freely available Web browser that was developed by students and staffs at NSF-supported National Center for Supercomputing Applications (NCSA) at the University of Illinois. NCSA Mosaic Web browser allowed Web pages to include both graphics and text, and in less than 18 months, it became the Web "browser of choice" for more than a million users. Mosaic set off an exponential growth in the number of Web servers as well as the number of Web surfers.

Commercial firms also built their own networks after noting of the popularity and effectiveness of the growing Internet. In 1993, the proliferation of networks from the private suppliers led to an NSF solicitation that later outlined a new Internet architecture, which largely remains in place today. In 1995, NSF awarded contracts based on the solicitation agreement with the private suppliers for three network access points, to provide connection points between commercial networks, and one routing arbiter, to ensure an orderly exchange of traffic across the Internet. In addition to that, NSF also signed a cooperative agreement to establish the next generation of very high performance Backbone Network Service that was complimented with decommissioning of the NSFNET backbone in April 1995.

After that, the road to a self-governing and commercially viable Internet started with a period of remarkable growth. The year 1998 marked as the end of NSF's direct role in the Internet, and the network access points and routing arbiter functions were transferred to the commercial sectors.

REFERENCES

1. Glowniak, J. (1998). History, structure, and function of the Internet. *Seminars in Nuclear Medicine,* 28(2), 135–144. https://doi.org/10.1016/s0001-2998(98)80003-2.
2. Licklider, J. C. R. (1963, April). Intergalactic Computer Network. ARPA.
3. Cohen-Almagor, R. (2013). Internet history. *Moral, Ethical, and Social Dilemmas in the Age of Technology: Theories and Practice*, 19–39. IGI Global. https://doi.org/10.4018/978-1-4666-2931-8.ch002.
4. Leiner, B. M., Cerf, V. G., Clark, D. D., Kahn, R. E., Kleinrock, L., Lynch, D. C., … & Wolff, S. (2009). A brief history of the Internet. *ACM SIGCOMM Computer Communication Review*, 39(5), 22–31.
5. Kleinrock, L. (2010). An early history of the Internet [history of communications]. *IEEE Communications Magazine*, 48(8), 26–36.
6. Bay, M. (2019). Conversation with a pioneer: Larry Roberts on how he led the design and construction of the ARPANET. *Internet Histories*, 3(1), 68–80.
7. Campbell-Kelly, M., & Garcia-Swartz, D. D. (2013). The history of the internet: the missing narratives. *Journal of Information Technology*, 28(1), 18–33.
8. Odlyzko, A. (2000). The history of communications and its implications for the Internet. Available at SSRN 235284.
9. Leiner, B. M., Cerf, V. G., Clark, D. D., Kahn, R. E., Kleinrock, L., Lynch, D. C., … & Wolff, S. S. (1997). The past and future history of the Internet. *Communications of the ACM*, 40(2), 102–108.
10. Crocker, S. D. (2021). Arpanet and its evolution—A report card. *IEEE Communications Magazine*, 59(12), 118–124.
11. Roberts, L. (1988). The Arpanet and computer networks. *A History of Personal Workstations*, 141–172.
12. Marill, T., & Roberts, L. G. (1966, November). Toward a cooperative network of time-shared computers. *Proceedings of the November 7–10, 1966, Fall Joint Computer Conference*, 425–431. https://doi.org/10.1145/1464291.1464336.
13. Roberts, L. G., & Wessler, B. D. (1970, May). Computer network development to achieve resource sharing. *Proceedings of the May 5–7, 1970, Spring Joint Computer Conference*, 543–549. https://doi.org/10.1145/1476936.1477020
14. Cerf, V., & Kahn, R. (1974). A protocol for packet network intercommunication. *IEEE Transactions on Communications*, 22(5), 637–648.
15. Baran, P. (1964). On distributed communications networks. *IEEE Transactions on Communications Systems*, 12(1), 1–9.
16. Roberts, L. G. (1978). The evolution of packet switching. *Proceedings of the IEEE*, 66(-11), 1307–1313.
17. Kita, C. I. (2003). JCR Licklider's Vision for the IPTO. *IEEE Annals of the History of Computing*, 25(3), 62–77.
18. Roberts, L. G. (1967, January). Multiple computer networks and intercomputer communication. *Proceedings of the First ACM Symposium on Operating System Principles*, 3.1–3.6. https://doi.org/10.1145/800001.811680.
19. Ida, N. (2022). History of communication and the Internet. *Handbook of Nondestructive Evaluation 4.0*, 77–93. Cham: Springer International Publishing.

20. Wheen, A. (2010). The birth of the Internet. *Dot-Dash to Dot. Com*, 127–138. New York: Springer.
21. Hauben, M., & Hauben, R. (2006). Behind the Net: The Untold History of the ARPANET and Computer Science. Netizens: On the History and Impact of Usenet and the Internet. California: IEEE Computer Society Press.
22. Strawn, G. (2021). Masterminds of the NSFnet: Jennings, Wolff, and Van Houweling. *IT Professional*, 23(6), 67–69.
23. Rogers, J. D. (1998). Internetworking and the politics of science: NSFNET in Internet history. *The Information Society*, 14(3), 213–228.
24. Berners-Lee, T. J. (1992). The world-wide web. *Computer Networks and ISDN Systems*, 25(4–5), 454–459.

4 Ethernet

ABBREVIATIONS

CATV	Community Access Television
CSMA\CD	Carrier Sense Multiple Access with Collision Detection
IEEE	The Institute of Electrical and Electronic Engineers
IP	Internet Protocol
LAN	Local Area Network
LLC	Logical Link Control
MAC	Medium Access Control
PoE	Power over Ethernet
SFD	Start frame delimiter
TCU	Terminal Control Unit
UH	University of Hawaii
UHF	Ultra High Frequency
UTP	Unshielded twisted pair

4.1 INTRODUCTION

"Ethernet" is the most common and popular access technology available today and is by far the most widely used local area networking (LAN) technology. Ethernet has gained so much importance in networking that it is worth mentioning it as an industry standard for packet-based computer networking technology for LANs. It is considered the foundation of most wired communication network and has become a network of choice to get the wired connection for homes and enterprises around the globe. In this chapter, we are going to learn about the concept of "Ethernet" as a historical journey and further analyze its impact on computer networking.

Provision of network access speed by the Ethernet and innovations in the communication technology has been the major confidence booster for its commercial development, and the development was supported by many industries. With the success of Internet, the growth in Internet Protocol (IP) to carry data, voice, and video continued to drive the demand for greater network bandwidth. Gigabit Ethernet and Ten Gigabit Ethernet are providing the core of data center computing and storage resources. Ethernet innovation was therefore another important milestone in the history of computer network parallel to the development of Internet, and mutually, they both benefitted the overall network experience. Internet gave the concept of packet-oriented data transfer in a decentralized network environment and gave the idea of network of networks. Internet gave that core network platform with global reach that can be integrated with many and multiple types of organizational networks whether small or large they may be. Ethernet as a networking technology gave the speed of the local network or good usable bandwidth to the network data processing.

DOI: 10.1201/9781003302902-4

Ethernet became such a technology that every organization hoped for it to have it within their organization, irrespective of the organization having the connection to Internet or not. Ethernet started a new industrial growth in that area of network hardware manufacturing, which was well supported with major industrial collaboration on the baseline of standardization of Ethernet.

4.2 ALOHA

The ALOHA systems network also famous as ALOHANET is a computer networking system that was developed by the University of Hawaii (UH) and is considered as the ancestor of all shared media networks [1, 2]. ALOHA is considered one of the early computer network designs, its development started at the University of Hawaii under the leadership of Dr. Norman Abramson, Dr. Franklin Kuo, and a team of faculty and students in September 1968. At UH, they started working on a research program for investigating the use of radio communications for computer-computer and console-computer links. They further mentioned in the report about a remote-access computer system as the ALOHA system-under development as part of the research program. They discussed about the advantages of the radio communications over conventional wired communication for interactive users of a large computer network. That research program on ALOHA SYSTEMS at UH was composed of a large number of research projects, but they gave more focus in that report on a novel form of random-access radio communications.

The ALOHA SYSTEMS planned for establishing a network with linking of interactive computer users and remote-access input-output devices away from the UH main campus to the central computer with UHF radio communication channels. University's main campus computer center was operating with an IBM 360/65, and many other smaller machines at Honolulu and was supporting many colleges on the islands of Oahu, Kauai, Maui, and Hawaii, a number of research institutes expanded across the State along.

4.2.1 The ALOHA SYSTEM – Architecture

The ALOHA SYSTEM had central computer IBM 360/65 linked to the radio communication channel through a small interface computer. The design of this multiplexer was mostly based on the design of IMPs used in ARPA computer network. Abramson presented a paper based on the research work conducted on ALOHA SYSTEM with heading "THE ALOHA SYSTEM – Another alternative for computer communications" in 1970. Figure 4.1 is the architecture diagram based on Abramson's ALOHA SYSTEM.

MENEHUENE, an HP 2115A version of computer with 16-bit word size, a cycle time of 2 μs and an 8K word core capacity, was the Hawaiian version of the ARPA's IMP, with consideration of the radio communication and some other differences. The ALOHA SYSTEM was also to link with the remote-access input-output devices and small satellite computers through MENEHUNE [3].

The ALOHA SYSTEM was assigned with two 100 KHz channels at 407.350 and 413.475 MHz, with one of these channels assigned for data from MENEHUENE to the remote consoles and the other one for the data from remote consoles to the

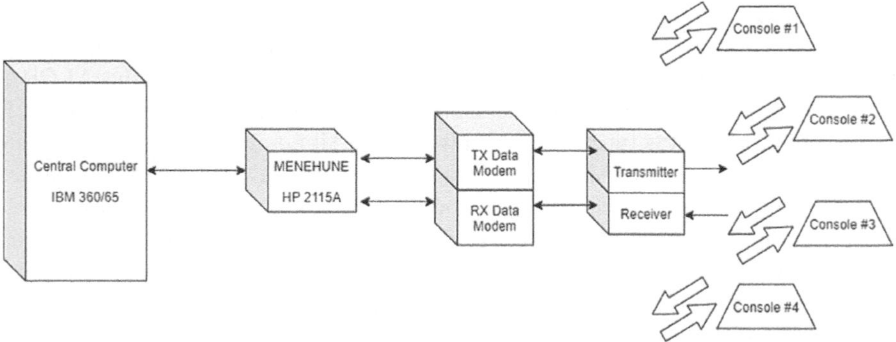

FIGURE 4.1 The ALOHA SYSTEM architecture.

MENEHUENE. Each of these channels was supposed to operate at the rate of 24,000 baud. The messages from the MENEHUNE to remote consoles had no problem due to the buffering provided by the MENEHUENE at the computer center, however the messages from the remote consoles to the MENEHUENE had to be ordered in queue and according to some priority scheme and to transmit them sequentially.

In the ALOHA SYSTEM, as was proposed by Abramson, a method of random-access communication which allows each console to use a common high-speed data channel without the necessity of central control or synchronization. The messages to and from the MENEHUENE were to be transmitted in the form of packets, where each packet will have a fixed length of 80×8 bit characters along with 32 bits for identification and control and 32 parity bits. Therefore, a packet of the ALOHA SYSTEM consist of 704 bits totally and to last for 29 milliseconds at a data rate of 24,000 baud. The use of parity bits in each of the packet will be for the error detection purpose.

Each user from the remote console of the ALOHA SYSTEM was supposed to transmit the data packets over the same high data rate channel to the MENEHUENE in a completely unsynchronized manner. The packet received by the MENEHUENE was to be recognized if and only if it was received without any error. The transmitting console has to wait till a certain amount of time for the acknowledgment on the transmitted packet back from the MENEHUENE; if not received, then it has to retransmit it again until the successful transmission and acknowledgment occurred.

A packet can be incorrectly received in two separate conditions: (a) it can be received incorrectly due to random noise error in the channel, and (b) errors caused due to interference with another packet transmitted over the channel. It was observed that the error due to interference can be of much importance in case of a large number of users in the channel and will limit the number of users and the amount of data over the random-access channel.

4.2.2 ALOHANET

As we discussed, the ALOHA SYSTEM was started by the UH as a research project to search for an alternative networking system to the telephonic system in 1968. It was the need of the time to search for an alternative network as it became apparent that the existing telephone network architecture was not suitable for rapidly emerging data networking needs of the 1970s. One of the goals of the ALOHA SYSTEM was to investigate about the use of radio communication in computer network. The research was further followed with some significant efforts in this direction. The ALOHA SYSTEM developed to illustrate general principles for a packet broadcasting network, which discussed about the relationship between information theory and the design of real information system. The ALOHANET [2,4–6] was an experimental network designed on the basis of the general design principles as were given in the ALOHA SYSTEM. The development of this experimental network ALOHANET only happened with a matter of time as it needed the confidence building and appreciation from the research community. As one of the major architectural differences was traditional telephone network's point-to-point telephone channel, the radio access packet network, however, was with broadcasting or multiple access radio channels.

The first ALOHA packet broadcasting unit of ALOHANET became operational in June 1971. That had formatting of ALOHA packet and packet retransmission protocols accomplished by a special purpose equipment called as the Terminal Control Unit (TCU). ALOHANET had used RS232 interface to connect a terminal with the TCU. A user was connected with the central system at a data rate of 9,600 bits/sec anywhere within the range of about 100 km radio coverage of ALOHANET. In 1971–1972, ALOHANET got equipped with many additional TCUs and the network came into existence. It also became clear by then that the key differentiator of the ALOHANET was not the use of radio communications for computers, but the use of broadcast communication architecture for the radio channel. The launch of Westar I satellite by the US in 1974 increased the possibility of using satellite channel as the broadcast channel and the use of ALOHA channels in satellite systems.

It was later decided to obtain a satellite link for ALOHA system in Hawaii to connect it with the expanding ARPANET packet switching network on the US mainland. By that time, the ARPA took over the funding of ALOHA system under the direction of Dr. Lawrence Roberts. ARPA had installed one of its IMPs at Hawaii location for a satellite connection with the ALOHA system on December 17, 1972. ALOHANET had the first commercial satellite link that used a single satellite voice channel to transmit data at 56 Kbits/sec. The packet broadcast was operational in the ALOHA system using an ATS-1 satellite in 1973.

With the innovation of microprocessors by 1974, the logic of TCU was supposed to be taken over by the microprocessors and several new terminal controllers based on Intel 8080. The network protocols were now to be implemented in the software and that can be modified for all operating units, and these new types of controllers were called as the programmable control units.

It took more time for the commercial use of packet broadcasting in UHF radio-based system as was done in the experimental ALOHANET due to the regulatory constraints of the frequency assignments. Motorola announced the introduction of unslotted ALOHA channel in UHF band at the data rate of 4,800 bits/sec in its PCX personal computer.

All the wireless communication happening today, including mobile communication, satellite communication, cellular, and Wi-Fi, utilizes the ALOHA protocol to establish the initial link.

4.2.3 Pure and Slotted: ALOHA Protocols

Both of the protocols, Pure and Slotted ALOHA [7], are the Random-Access Protocols, implemented on the medium access control (MAC) sub-layer of the Data Link Layer. ALOHA protocol is used to determine which of the competing station should get the next chance to access the multi-access channel at MAC layer.

The Pure ALOHA allows its stations to transmit the data whenever the data is available to transmit at arbitrary times. There is always a possibility of the collision of data frames if every station transmits the data without checking channel acquisition whether the channel is free or not. For successful transmission, the acknowledgment has to arrive for the transmitted frame if received successfully without any error from collision or due to noise in the channel. Otherwise, the frame may collide or be damaged due to simultaneous transmission of frames in the channel, which will get discarded and to be retransmitted after waiting for a random amount of time till the frame will be transmitted successfully.

Slotted ALOHA was introduced as another method to improve the capacity of the Pure ALOHA. In the Slotted ALOHA, it was proposed to divide the time into certain discrete intervals that were called time slots where each of the time slots corresponds to the length of the frame. As compared with Pure ALOHA, Slotted ALOHA does not allow transmission of the data frame whenever the station has the data to send. In the Slotted ALOHA, every station has to wait till the next time slot begins, and the Slotted ALOHA permits only the transmission of each data frame in the new time slot.

4.3 ETHERNET HISTORY

Historically, the design of Ethernet started with the basic idea of the packet collision and packet retransmission, which was developed in ALOHA network. UH's ALOHA is considered as the ancestors of all shared media networks. ALOHA network demonstrated it first that the communication channels can be shared on a large scale effectively and efficiently using simple random-access protocols. It had provided public demonstration of a wireless packet data network, which led the development of Ethernet and personal wireless communication technologies.

Unlike an ALOHA network which is a star network with an outgoing broadcast channel and an incoming multi-access channel, an Ethernet supports many-to-many communication with a single broadcast multi-access channel.

Robert Metcalfe and David Boggs at Xerox Corporation, Palo Alto Research Center, CA created world's first LAN applying the principles of the ALOHA network, in 1973. They called it initially as the ALTO ALOHA and later changed the name to Ethernet. This was the first version of Ethernet that ran at a speed of up to 2.94 Mbits/s. Unfortunately, this was not well commercialized except the use of Ethernet at the White House.

4.3.1 Experimental Ethernet

Metcalfe started working on some similar access mechanism considering his inspiration from the ALOHA system where the access through a shared communications channel [8]. As was the ALOHA channel in the ALOHA SYSTEM, his idea of the shared communication channel was something that he mentioned as Ether based on the concept of luminiferous Ether. The term Ether was once used to represent an invisible medium through which electromagnetic radiations were supposed to propagate. An Ethernet transmitter can broadcast packets as completely addressed transmitter bit sequences into the Ether similar to an ALOHA radio transmitter. Metcalfe mentioned Ether in the network as a logically passive medium that will be used to propagate digital signals and can be constructed by using any media types like coaxial cables, twisted pairs, and optical fibers.

Metcalfe along with his colleagues at Xerox PARC created the first Experimental Ethernet in 1972. That Experimental Ethernet [8] was used for networking between different Xerox Alto computer systems, servers, and laser printers. This Ethernet was with the data transmission rate of 2.94 Mbit/s. Initially, the Experimental Ethernet was named as the Alto ALOHA network. However, Metcalfe changed the name of this network to Ethernet in 1973, as he wanted to make this more generic networking system that can support any computer not specifically Alto computer systems only (Alto was a Xerox personal workstation with graphical user interface) and also with an intention to point out that this networking system had evolved well beyond the ALOHA system.

Figure 4.2 was drawn by Metcalfe for describing Ethernet and its components used to present Ethernet at National Computer Conference in June 1976. Four major components are the Ether, transceivers, interfaces, and controllers. Controller is the low-level software or firmware component specific to every station. Controller is used to generate the retransmission interval dynamically for a collision detected by the source station based on the collision count. Interface is the hardware component on every station and is used to serialize and de-serialize the parallel data used by its station. Low-loss coaxial cable was used as Ether with off-the-shelf CATV taps and connectors. The experimental network was supposed to work for 1 km of cable length with data transmission capacity of the cable up to peak value 3 Mbit/s and supported

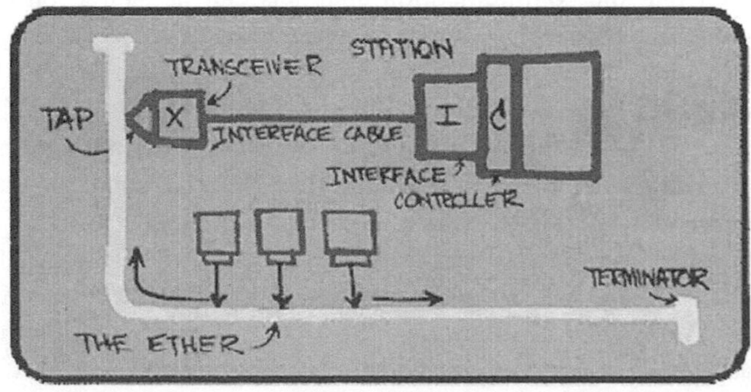

FIGURE 4.2 Ethernet diagram presented at National Computer Conference [8].

256 Ethernet stations on the network. Every station on the Ethernet is having a controller and an interface which is further using interface cable, transceiver, and tap to connect with the Ether. This setup was having the tap to establish a physical connection with the Ether. The transceiver was attached directly to the Ether, powered and controlled by 5 twisted pairs in an interface cable. The interface cable were carrying the different signals such as transmit data, receive data, interference detect, and the power supply voltages.

The topology of the Ethernet was of an un-rooted tree, and so the Ether can be extended from any point to any direction. It can branch at entrance of a building corridor and can avoid multipath interference. Ether should have only one path between any of its source and destination for smooth operation. Otherwise, in case of more than one path, the transmission will interfere with itself, which will make the repeated arrival of the data frame at its intended destination through multiple path lengths. Stations wishing to join the Ethernet have to tap the Ether at the nearest suitable point. Interconnects are having central control with respect to Ether and the distributed control of individual stations.

The Ether was to be shared in a controlled manner among its stations such that no two stations can transmit data packet at the same time. At the time of detecting a collision, the station has to abandon the attempt and retransmit the packet after some random time duration. At the time of no other transmission taking place, the transmitting station can send a packet at will and receiver will receive it with error. Packet interference will increase with the number of transmission attempts increasing by the stations. Each of the stations is equipped with an Ethernet controller to adjust the mean retransmission interval proportional to the frequency of collision.

4.3.1.1 Error Detection Mechanism

Carrier Sense Multiple Access with Collision Detection (CSMA/CD) mechanism was developed and included in the Ethernet. It consists of two concepts combined together: carrier sense of stations for the presence of a packet due to carrier transition in Ether, and detection of collision due to simultaneous transmission of two packets from two stations. As the packet placed on the Ether by the station, the bits get phase encoded and the presence can be felt with the transition present in the Ether. Stations can listen to this transition as the packet passes through the Ether and therefore can be detected. The station can sense the carrier of the passing packet and can delay sending of its own packet. This feature of carrier sense was not present in the past developed ALOHA network.

With carrier sense enabled, the collision can happen only when two or more stations begin transmitting packet simultaneously when they find the Ether to be idle or unused. For the detection of interference due to collision of the packets, each of the transceiver was equipped with an interference detector. Transceiver can detect interference by observing the difference in the bit value it was attempting to send and comparing with the bit value receiving from the Ether. With the collision detection feature, the station can know quickly about the packet got damaged without waiting for a long acknowledgment timeout. The retransmission of the packet will be scheduled quickly the moment collision detected in the Ether. The retransmission interval can be adjusted to optimize the channel efficiency based on the frequency

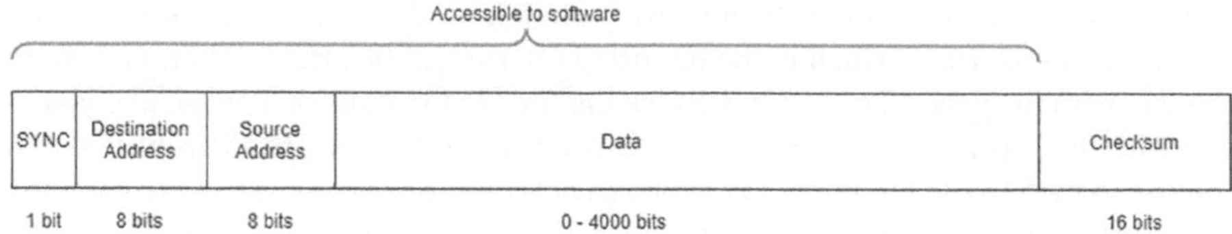

FIGURE 4.3 Diagram representing Ethernet packet format.

of detected interference. This is further supported by the mechanism of collision consensus enforcement, with that a station determining a collision will temporarily jam the Ether so that all the participating stations to get informed about the collision and abort transmission [9].

To detect the error in the packet received by the station, a checksum is recalculated and to be compared with the checksum received from the packet. Packet will be discarded if the checksum recalculated is not consistent with the checksum received. This will help to minimize the transmission errors, impulse noise errors, and other errors due to undetected interference.

Every Ethernet packet should have a source address and destination address information in the packet's header as shown in Figure 4.3. Every packet has been allocated 8-bit for the destination address field, followed by next 8-bit for source address to accommodate with 256 stations on the Ethernet.

4.3.2 Ethernet Developments in History

- **1973**: Experimental Ethernet of 2.94 Mbits/s on coaxial cable bus created by Robert Metcalfe and David Boggs at Xerox Corporation with single byte node address unique only to individual network.
- **1980**: DIX released the first commercial Ethernet specification as Ethernet Version-1 under the name of Ethernet Blue Book or DIX standard, commonly referred as Ethernet DIX80 and defined Thick Ethernet of 10 Mbps CSMA/CD.
- **1980**: IEEE Project 802 was formed for standardization of LAN technology. IEEE formed working group for this standardization and reworked on some of the parts of DIX Ethernet, with regard to the definition of frame.
- **1980**: IEEE finalized overall packet specifications considering the Ethernet standard, suitability of the transmission medium to be agreed upon. SynOptics Communications helped with a low-cost transmission medium by developing a method to transmit 10Mbps Ethernet signals over twisted-pair cables in the late 80s. This combination of low-cost transmission medium and the packet technology specifications led to the wide deployment of Ethernet.
- **1982**: DIX released the second commercial version as Ethernet Version-2 also referred to as Ethernet DIX82; i.e. still the standard Ethernet technology is in use today. Ethernet controllers based on DIX standard were commercially available for the first time from 1982.

- **1983**: First IEEE standard Ethernet technology launched and developed by the 802.3 working group of the IEEE 802 committee that defined access method and physical layer specifications. This was named IEEE 802.3 standard based on Carrier Sense Multiple Access with Collision Detection (CSMA/CD) process, which included IEEE 802.2 Logical Link Control (LLC).
- **1983**: Novell released Netware'86 software using proprietary frame format on IEEE 802.3 standard preliminary specification; i.e. still used today to manage printers and servers.
- **1985**: IEEE standard IEEE 802.3a was approved to define standard for Thin Ethernet, 10Base2 10 Mbits/s (1.25 MB/s) over thin coaxial cable, is also called as cheapernet. This Ethernet standard was dominated and was in demand after its launch. This has 10 representing for 10 Mbits/sec of data transmission speed, Base stands for baseband signaling and 2 for the maximum segment length of the cable about 200 m.
- **1985**: IEEE 802.b standard with 10Broad36 networking capability to support 10 Mbit/s Ethernet signals on standard 75 Ω cable television cable (CATV) over a distance of 3,600 m. 10Broad36 modulates data onto a higher frequency carrier signal, much similar to an audio signal that would modulate a carrier signal transmitted in a radio station. Unfortunately, this Ethernet standard was less successful than its contemporaries because of the high equipment cost and complexity associated.
- **1987**: IEEE 802.d standard 10BFOIRL with Fiber Optic Inter-Repeater Link, a specification of Ethernet over optical fiber. This standard was designed as a back-to-back transport between repeater hubs for decreasing latency and time of collision detection. This had two optical fiber cables that extended the overall distance between 10 Mbps repeaters up to 1,000 m.
- **1987**: IEEE 802.e standard or "1BASE5" or "StarLAN" as the first IEEE 802.3 standard for Ethernet over twisted-pair wiring. It was named as StarLAN by IEEE because it used a star topology from a central hub in contrast to the bus network of the shared cable 10BASE5 and 10BASE2 networks based on ALOHANET. One of the design goals of StarLAN was to reduce the Ethernet installation costs by reusing existing on-premise telephone wiring, and its compatibility with analog/digital telephone signals in the same cable bundle. The name also refers 1Base5 for the fact that 1Mbit/s was the original StarLAN speed and 5 representing the maximum cable segment length that depending on the cable performance up to 500 m was possible.
- **1988**: AT&T released StarLAN 10 at transmission speed at 10 Mbit/s (StarLAN 1 was operational at 1 Mbits/s). AT&T StarLAN 10 and SynOptics LattisNet provided the basis for the 10 Mbit/s Ethernet standard 10Base-T.
- **1990**: IEEE 802.3i standard over twisted-pair specification as 10Base-T. This was the popular Ethernet standard that became the preferred media type for Ethernet with providing transmission speed of 10Mbits/s over UTP category 3 cable.
- **1993**: IEEE 802.3j standard with specification 10Base-F to provide 10 Mbit/s (1.25 MB/s) over Fiber Optics. Several variants of this Ethernet

standard were available as 10Base-FL, 10Base-FB and 10Base-FP under the 10Base-F specification. The maximum segment cable length possible with this specification was about 2,000 m.

- **1995**: IEEE introduced IEEE 802.3u standards with 100Base-TX, 100Base-T4, and 100Base-FX as the Fast Ethernet and was the fastest among the versions available of Ethernet remained until the introduction of Gigabit Ethernet. In the Fast Ethernet standard, the physical layer carry data at the transmission speed of 100 Mbit/s. Before Fast Ethernet the prior speed available with Ethernet was of 10 Mbit/s. Among all the Fast Ethernet variants, 100Base-TX is the most common. Fast Ethernet is in itself an extension of the 10 Mbits/s Ethernet standard that runs over twisted-pair or optical fiber cable. It may have a star wired bus topology, similar to the IEEE 802.3i 10Base-T, and its devices are considered as backward compatible generally with existing 10Base-T systems. That can enable plug-and-play upgrades from 10Base-T.
- **1997**: IEEE 802.3x standard with full-duplex and the first flow control mechanism for Ethernet was defined with introducing the pause frame. Ethernet flow control mechanism used for temporarily stopping the transmission of data on Ethernet family networks to avoid packet loss due to congestion in the network.
- **1997**: IEEE 802.3y standard specification for 100Base-T2 with data transmission speed of 100 Mbit/s (12.5 MB/s) over voice-grade twisted pair. This supports data transmission over two copper pairs of category 3 cable, where data can be simultaneously transmitted and received on both of the pair cables allowing full duplex.
- **1998**: IEEE 802.3z standard as 1000Base-X popularly known as Gigabit Ethernet providing transmission speed at 1 Gbit/s (125 MB/s) over Fiber Optic and was the replacement of Fast Ethernet in wired local networks because of the speed improvement over Fast Ethernet.
- **1999**: IEEE 802.3ab Ethernet standard also known as 1000Base-T defined data transmission over unshielded twisted pair (UTP) category 5, 5e or 6 cabling for Gigabit Ethernet. With this introduction, the 802.3ab became a desktop technology where the organizations could use their existing copper cabling infrastructure.
- **1998**: IEEE 802.3ac standard was introduced to allow Q-tag to include 802.1Q VLAN information and 802.1p priority information. This standard had maximum Ethernet frame size extended from 1,518 to 1,522 bytes to accommodate four byte of the VLAN tag.
- **2000**: IEEE 802.3ad standard was introduced for Link aggregation of parallel links Gigabit channel bonding.
- **2002**: IEEE 802.3ae standard to provide 10 Gigabit Ethernet over fiber through the specifications 10GBase-SR, 10GBase-LR, 10GBase-ER, 10GBase-SW, 10GBase-LW, 10GBase-EW for transmitting Ethernet frames at a data rate of 10 Gigabits/s. This Ethernet standard defined only full-duplex point-to-point links connected by network switches. This standard do not carry from the previous generations of Ethernet standards

the shared-medium CSMA/CD operation, thus half-duplex operation and repeater hubs do not exist in 10Gigabit Ethernet. Also encompasses a number of different physical layer standards respective to prior standards.

- **2003**: IEEE 802.3af Ethernet standard defined Power over Ethernet (PoE) providing up to 15.4 W of DC power (minimum 44 V DC and 350 mA) on each port to pass electric power along with data on twisted-pair Ethernet cable. It has a single cable to provide both data connection and electric power to the networking devices. Some of the PoE specification variants are alternative A, alternative B, and 4PPoE.
- **2004**: IEEE 802.3ah by adding two more Gigabit fiber standards which were the 1000BASE-LX10 and the 1000BASE-BX10. This version of Ethernet standard was part of larger group of protocols known as Ethernet in the First Mile. This standard refers to the use of one of the Ethernet family networking technologies between a telecommunications service provider and a customer's premises. From the end customer's point of view, it is their first mile toward the service provider's core network, although from the access network's point of view it is the last mile toward customer premise.
- **2004**: IEEE 802.3ak Ethernet standard as 10GBase-CX4 specification for providing 10 Gbit/s (1,250 MB/s) Ethernet data transmission rate over twin-axial cables. This was the first 10 Gigabit copper standard of Ethernet published by IEEE 802.3 working group.
- **2006**: IEEE 802.3an with specification of 10GBase-T to provide 10 Gbits/s (1,250 MB/s) Ethernet frames over unshielded twisted pair (UTP), and distances up to 100 m (330 ft).
- **2006**: IEEE 802.3aq Ethernet standard with specification of 10GBase-LRM (long reach multimode) 10 Gbit/s (1,250 MB/s) Ethernet over multimode fiber. This standard allows distance up to 220 meters (720 ft) on FDDI grade multimode fiber and maximum reach of 220 m on the fiber types OM1, OM2 and OM3.
- **2007**: IEEE 802.3ap Ethernet standard as the backplane Ethernet 1 and 10 Gbit/s (125 and 1,250 MB/s) over printed circuit boards.
- **2009**: IEEE 802.3at Ethernet standard for Power over Ethernet (PoE) specification enhancement up to 25.5 W from earlier 802.3af release of having 15.4 W.
- **2009**: IEEE 802.3av standard as 10G-EPON to provide 10 Gbit/s Ethernet Passive Optical Network standard. It is a type of passive optical network point-to-multipoint by using the passive optical fiber splitters rather than the powered devices for fan-out from hub to customers.
- **2010**: IEEE 802.3az standards for Energy-Efficient Ethernet as a set of enhancements to the twisted-pair and backplane Ethernet family to reduce power consumption in the network. With the intention of reducing power consumption about more than 50%, while retaining full compatibility of networking equipment. Green Ethernet was the technology developed by some companies to reduce the power requirement of Ethernet before the final IEEE standard was ratified in 2010. Therefore, the Green Ethernet technology can be considered as a superset of the IEEE 802.3az Ethernet standard.

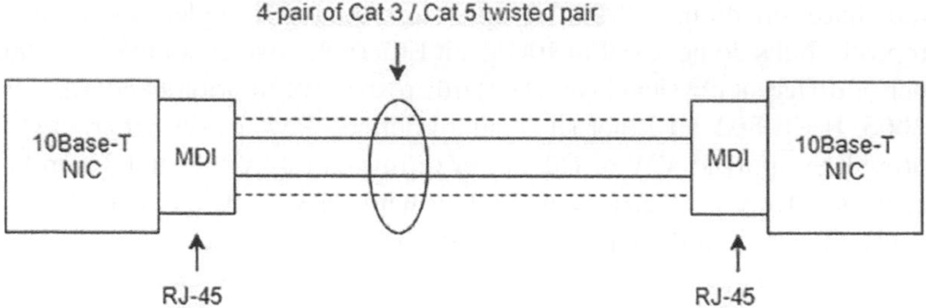

FIGURE 4.4 10Base-T Ethernet Link with four-pair cables, two-used and two-unused.

- **2010**: IEEE 802.3ba standard for extending the 802.3 Ethernet frames operational speeds up to 40 and 100 Gbit/s, for the first time two different Ethernet speeds were specified in a single Ethernet standard. With providing a significant increase in bandwidth while maintaining maximum compatibility with the installed base of 802.3 interfaces. Two different speeds were included to support the 40 Gbit/s rate for local server applications and the 100 Gbit/s rate for Internet backbones.
- **2011**: IEEE 802.3bd standard for priority-based flow control. It was an amendment of the IEEE 802.1 Data Center Bridging Task Group to develop an amendment for IEEE 802.3 standards for adding a MAC Control Frame to support priority-based flow control.
- **2013**: IEEE 802.3bk standard amendment to IEEE 802.3 standard to define the physical layer specifications and management parameters for EPON on point-to-multipoint passive optical networks. It was to support extended power budget classes of PX30, PX40, PRX40, and PR40 PMDs.
- **2015**: IEEE 802.3bm standard for supporting 100G/40G Ethernet data rate for optical fiber.
- **2016**: IEEE 802.3bn standard as specifications 10G-EPON and 10GPass-XR to provide passive optical networks over coaxial cables.
- **2016**: IEEE 802.3bq Ethernet standard with specification of 25G/40GBase-T to provide Ethernet speeds of 25G/40GB for four-pair balanced twisted-pair cabling with two connectors over 30 m distances.
- **2017**: IEEE 802.3bs standard to provide data rate of 200Gbit Ethernet (200 Gbit/s) over single-mode fiber and 400Gbit Ethernet (400 Gbit/s) over optical physical media.
- **2018**: IEEE 802.3bt as the specification for third generation of Power over Ethernet (PoE) providing up to 100 W of Ethernet power using all four pairs balanced twisted-pair cabling (4PPoE), that includes 10GBase-T, lower standby power and enhancements to support IoT applications.

4.4 ETHERNET CONCEPTS

The Ethernet networking technology comprises the hardware and the software systems working together to deliver the data has been through many evolutions over the time. Ethernet was earlier designed as a coaxial bus system initially and since then many alternatives have evolved to support it at the physical layer. By the IEEE 802 project committee, several of the important standards have been defined comprising of the evolution of the Ethernet, physical media types, power requirement, speed, etc. [10].

4.4.1 Physical Layers of Ethernet

Physical layer of Ethernet has been defined on the basis of certain electrical or bit rate specifications as per the IEEE 802.3 standards [11].

4.4.1.1 10 Mbit/s Ethernet

Original 10 Mbit/s Ethernet was based on coaxial cables of type "thick coax" with the approximate diameter of about half an inch. Another type of coaxial cable was of "thin coax" with diameter nearing about quarter of an inch. Coaxial cables have become obsolete now as the modern Ethernet are mostly based on twisted pairs and optical fiber cables.

10Base-T Ethernet system used twisted pair with star topology for Ethernet installations. 10Base-T Ethernet adaptors were using internal transceivers and RJ-45 connectors, with Manchester-encoded 10Mbps bit-serial communication over two twisted-pair cables. And, it was designed originally to support 10 Mbps Ethernet signals transmission over common telephone cable of category 3. Figure 4.4 represents a 10Base-T link with four pairs of category 3 or category 5 twisted-pair cables, where two pairs are used and two pairs are unused.

10Base-F series were launched as the optical fiber media Ethernet variant for 10Mbps Ethernet. These media systems were very popular at times and were sold and used for many years. Optical systems use the pulses of light to send the Ethernet signals. There were several advantages of using the optical fiber media over the twisted pairs. As the optical fiber cables can provide long distance with less attenuation, support higher-speed Ethernet systems, used as backbone cabling, provides noise immunity and electrical isolation. The optical fiber media standards are 10Base-FL, 10Base-FB, and 10Base-FP. 10Base-FL replaced the older FORL link segment and can support a fiber optical link segment up to 2,000 m. Manchester signal encoding was used by the 10Base-FL systems to send signal through the optical fiber media.

4.4.1.2 Fast Ethernet

After the 10 Mbps Ethernet, the next popular speed variant was the 100 Mbps Ethernet also famous as the "Fast Ethernet" systems. These media systems were first defined in the Ethernet standard 802.3u with 100Base-TX, 100Base-T4, and 100Base-FX as the Fast Ethernet in 1995. The effort for increasing the speed of Ethernet by a factor of 10 to twisted pair on 10Base-T resulted in the development of separate 100 Mbps twisted-pair media standards as 100Base-TX and 100Base-T4 in 1995, and 100Base-T2 in 1997. Each of the standards has been defined with a different encoding requirement and a different set of media-dependent sub-layers. 100Base-TX was using the 4B/5B encoding, supporting full-duplex operation, with two pairs of UTP category-5 cables or Type-1 shielded twisted pair. The encoding procedure 4B/5B is the same as the encoding procedure used by FDDI, with a minor adaptations to accommodate the Ethernet frame control. The system is using one pair to receive the data signals and the other to transmit the data signals. All of the connections today with 100Base-TX are using the RJ45 connectors built-in Ethernet interfaces on computers or switches. 100Base-T4 was using 8B/6T encoding, supporting only half-duplex operation with four pairs of UTP category-3 or better cables. 100Base-T2 was using the PAM5x5 encoding, supporting full-duplex operation with two pairs of UTP category-3 or better cables.

100Base-FX was the Fast Ethernet system over optical fiber media and provided all the advantages of the 10Base-FL systems. 100Base-FX systems were widely used for switch uplinks and are still in use, but mostly faster standards are preferred like 1 or 10 Gbps uplinks for higher performance. Block-encoded signaling used for 100Base-FX which was originally developed for ANSI X3T9.5 FDDI standard. The standard requires two strands of multimode fiber optical cable per link, one for transmitting data, another for receiving of the data.

4.4.1.3 Gigabit Ethernet

1000 Mbps Ethernet, also known as the Gigabit Ethernet, was a remarkable achievement considering the speed of Ethernet. One thousand Mbps or Gigabit Ethernet is used by IEEE to describe this Ethernet media systems operated over twisted pair as well as over the optical fiber cable. 1000Base-T was the twisted-pair physical media standards using UTP copper cable. The signaling and encoding schemes used for 1000Base-T were the combination and extension of the encoding developed earlier for 100Base-TX, 100Base-T2, and 100Base-T4 media standards.

1000Base-X standard is used to represent the optical fiber media systems of the Gigabit Ethernet, was developed first in the 802.3z supplement to the IEEE standard. Gigabit Ethernet variants over optical fiber media system are available as 1000Base-SX, 1000Base-LX, and 1000Base-CX. Light pulses with short wavelength were used for the Ethernet signals in 1000Base-SX over multimode fiber. This system was using short distance lasers designed for connecting short length of optical fiber segments; therefore, the 1000Base-SX is preferably used inside the building to connect high performance servers or workstations. Light pulses with long wavelength were used for the Ethernet signals in 1000Base-LX over multimode or single-mode fiber, whereas the third-type 1000Base-CX was using short copper jumper, but it was not adopted for the market place and hence was not made available. The 1000Base-X switch interface can support both variants of the fiber standards: 1000Base-SX and 1000Base-LX. Both of these fiber Gigabit Ethernet variants require two strands of cable: one for transmitting and the other for receiving the data.

4.4.1.4 Ten Gigabit Ethernet

10 Gigabit Ethernet represents the family of Ethernet systems that can provide maximum data transmission speed up to 10 GB/sec. This system is also popular as 10GE, 10GbE, or 10 GigE, was defined under 802.3ae IEEE standard in 2002 that defined the basic 10 Gigabit Ethernet system with a set of optical fiber media standard. Gigabit Ethernet standard supports only full-duplex mode of communication. Several standards available are the 10GBase-R, 10GBase-X, 10GBase-T, and 10GBase-W.

10GBase-R standard is using 64B/66B signal encoding over the optical fiber media system with Ethernet specifications as 10GBase-SR, 10GBase-LR, 10GBase-ER, and 10GBase-LRM. 10GBase-X is based on 8B/10B signal encoding with specifications; 10GBase-LX4 over optical fiber media systems and 10GBase-CX4 over copper media systems.

10GBase-T Ethernet system to provide 10 Gigabit Ethernet speed over twisted-pair media systems, using 64B/65B signal encoding. It was approved by IEEE in the supplements to 802.an in 2006, 4 years after the initial 10Gigabit Ethernet.

4.4.2 Medium Access Control

The data link layer of the OSI model is composed of medium access control (MAC) sub-layer and the logical link control (LLC) sub-layer [12,13]. The MAC sub-layer controls the hardware responsible for interaction with the transmission medium. The LLC provides flow control and multiplexing for the logical link, whereas the MAC provides flow control and multiplexing for the transmission medium.

Considering the example of 10Mbps Ethernet system with several stations, working on half-duplex mode of operation. The MAC protocol works in the way, as each Ethernet equipped device operates independently of all other stations on the network without needing a central controller. Stations are operating in half-duplex mode and attached to the Ethernet cable connected through a signaling channel, also requires the CSMA/CD mechanism to control access in the shared medium. CSMA/CD protocol refers to the set of rules designed to arbitrate access through a shared channel among several Ethernet stations connected to that channel. Please be informed that today most of the Ethernet stations are using full-duplex mode of operation, which has a dedicated channel between the station and a switch port and hence no need to control access to the link.

Any station on the Ethernet is using broadcast delivery mechanism, where each frame transmitted over the channel is to be heard by every station. This mechanism is supported by keeping of the address-matching logic on every stations interface and helping the physical medium to be kept in much simpler. To send data over shared channel, a station has to first listen to the channel. When a station on the Ethernet is willing to transmit, it waits until it senses any carrier indicating that some other station is transmitting for the moment. As soon as it detects silence or found the channel as idle, the station forms an Ethernet frame and transmit it over the channel. Each frame sent over the shared channel is read by all Ethernet interfaces connected to the channel. While reading the bits of the signal, the interfaces look up at the field of the frame containing the destination address. Interfaces compare the destination address of the frame with its own 48 bit MAC address, and if an Ethernet interface's address matches the destination address in the frame, then it will have to process that frame. That interface will continue reading the entire frame and deliver the frame to the networking application on that computer. All other network interfaces will stop reading the frame after finding that the destination address does not match with their own address. All stations on the shared half-duplex network channel having traffic to send over the channel must contend equally for the next frame transmission opportunity. This has to happen with ensuring fair access to the shared channel with no single station to lockout the others from transmission over the channel. A MAC algorithm is embedded in the Ethernet interface of each station to facilitate fair access to the shared channel.

Multicasting is an efficient Ethernet delivery mechanism while sending the same Ethernet frame to multiple recipients of Ethernet. The same Ethernet frame can be received by a group of stations by using a multicast addressing. This will require a multicast group to be configured with a specific multicast address for a set of stations to receive particular frame. Data sent with the multicast address as the destination address in the frame will be received by all stations in the multicast group.

The broadcast address is a special case multicast address that consists of all ones in the 48-bit address. All Ethernet interfaces looking a frame with the broadcast address as the destination address will read the frame and process it.

4.4.3 Ethernet Frame

Ethernet frame [11] is one of the most important parameters of the Ethernet system. Signals in the form of Ethernet frames are transmitted in the network hardware including the Ethernet interfaces and transmission media to move the Ethernet data between computers or stations. Devices connected to the Ethernet should have an Ethernet interface, and are called as Ethernet station which may be a desktop, a computer, a printer, or any other device. Ethernet has a specific frame format with the bits formed up in specified fields, as shown in Figure 4.5.

Preamble: 7 Bytes of Preamble is used in the start of the Ethernet frame that indicates start of the frame. Preamble allow sender and receiver to establish bit synchronization. In the original 10 Mbps Ethernet system, it was provided to give the hardware and electronics some start-up time to recognize that a frame is being transmitted and alerting it to start receiving of the data. Therefore in the initial Ethernet systems Preamble was introduced to protect from the loss of frame bits due to signal delays, indicating the receiver to lock onto the data stream before the actual frame begins. But today's high-speed Ethernet systems do not need Preamble and use constant signaling that avoids the need for a preamble.

SFD: Start frame delimiter (SFD) is a 1-Byte field, which is designed to break the bit pattern of the preamble and signal the start of the actual frame. SFD is used to indicate that upcoming bits are starting of the frame, which is the destination address. Sometimes SFD is also considered as the part of Preamble with the size of 8 Bytes of Preamble.

7 byte	1 byte	6 byte	6 byte	2 byte	46 - 1500 byte	4 byte
Preamble	SFD	Destination Address	Source Address	Type/ Length	Data	CRC

FIGURE 4.5 Ethernet frame format.

Destination Address: Destination Address is a 6-Byte field, which contains the MAC address of destination interface connecting the station for which data is destined. The 6-Byte address is often called the hardware or physical address, pointing the address assigned to the Ethernet interface of the station. The hardware address is also called as the media access control (MAC) address, because of the reason that the Ethernet media access control system includes the frame and its addressing. Out of the 6 Bytes of the MAC address, the first 24 bits are called as organizationally unique identifier used by the Organization that build Ethernet interfaces. Then the next 24 bits are used to ensure that every MAC address is unique.

Source Address: Source address is also a 6-Byte field, which contains the MAC address of Ethernet interface of the source station.

Type/Length: Length is a 2-Byte field indicating about the length information of entire Ethernet frame. The maximum value of length can be 1,500 as per the limitations of Ethernet. This field is also most of the times used to identify the type of high-level network protocol contained in the data field.

Data: This is that field of the frame where actual data of Ethernet signals is placed. This field is also known as the Payload. IP packet containing the IP header and data will be inserted here if Internet Protocol is used over Ethernet, and the process is called as encapsulation of IP packet. The maximum possible data present may be 1,500 Bytes, and the minimum data is 46 Bytes. If the data length is less than the minimum required length of 46 Bytes, then padding with extra 0's is inserted to meet the minimum length requirement.

CRC: Cyclic Redundancy Check (CRC) is a 4-Byte field, which contains a 32 bits hash code of data which is also called as frame check sequence or checksum. The checksum is calculated over the Destination Address, Source Address, Length, and Data fields, and resulted checksum is attached in this field. After the frame is received at the destination, again a calculation of checksum is to be made at the destination; if the checksum calculated at the destination is not the same as sent checksum value, data received is corrupted and to be discarded.

4.5 CONCLUSION

Starting our discussion on the multi-access radio network that was popularly known as ALOHANET, we have further discussed about the evolution of Ethernet in this chapter. Going through the basic functionality of Ethernet, we have understood about the Ethernet frame format that is an important parameter of the Ethernet system. Ethernet provides a broadcast, multi-access environment using the carrier sense and collision detection techniques for local area networking. Ethernet became very optimum and convenient to the evolving networking technology and found its utilization at home and business networks and a very dominant local area network architecture [14]. Ethernet is such a technology that is continuously evolving and adapting itself to the needs of the networking world. Ethernet is also addressing the requirements of both operators and end users to make the resulting technology cost-efficient, reliable, and operate in a plug-and-play manner.

REFERENCES

1. Abramson, N. (1977). The throughput of packet broadcasting channels. *IEEE Transactions on Communications*, 25(1), 117–128.
2. Kuo, F. F. (1995). The ALOHA system. *ACM SIGCOMM Computer Communication Review*, 25(1), 41–44.
3. Abramson, N. (1970). The ALOHA system: Another alternative for computer communications. *Proceedings of the November 17–19, 1970, Fall Joint Computer Conference*, 281–285.
4. Abramson, N. (1985). Development of the ALOHANET. *IEEE Transactions on Information Theory*, 31(2), 119–123.
5. Abramson, N. (2009). The ALOHANET—Surfing for wireless data. *IEEE Communication Magazines*, 47(12), 21–25.
6. Kuh, A., & Abramson, N. (2021). In memoriam: Norman Abramson. *IEEE Communications Magazine*, 59(1), 6–7.
7. Roberts, L. G. (1975). ALOHA packet system with and without slots and capture. *ACM SIGCOMM Computer Communication Review*, 5(2), 28–42.
8. Metcalfe, R. M., & Boggs, D. R. (1976). Ethernet: Distributed packet switching for local computer networks. *Communications of the ACM*, 19(7), 395–404.
9. Kleinrock, L., & Tobagi, F. (1975). Packet switching in radio channels: Part I-carrier sense multiple-access modes and their throughput-delay characteristics. *IEEE Transactions on Communications*, 23(12), 1400–1416.
10. Law, D., Dove, D., D'Ambrosia, J., Hajduczenia, M., Laubach, M., & Carlson, S. (2013). Evolution of Ethernet standards in the IEEE 802.3 working group. *IEEE Communications Magazine*, 51(8), 88–96.
11. Spurgeon, C. E. (2000). *Ethernet: The Definitive Guide*, Sebastopol, CA: O'Reilly Media, Inc.
12. Held, G. (2002). *Ethernet Networks: Design, Implementation, Operation, Management*, Hoboken, NJ: John Wiley & Sons.
13. Rom, R., & Sidi, M. (2012). *Multiple Access Protocols: Performance and Analysis*, Berlin, Heidelberg: Springer Science & Business Media.
14. Hellige, H. D. (1994). From SAGE via ARPANET to ETHERNET: Stages in computer communications concepts between 1950 and 1980. *History and Technology, an International Journal*, 11(1), 49–75.

5 Journey of Cables – From Coppers to Optical Fiber

ABBREVIATIONS

AWG	American wire gauge
ADSL	Asymmetrical digital subscriber line
CCTV	Closed-Circuit Television
EMI	Electromagnetic interference
FDM	Frequency-division multiplexing
FEP	Fluorinated Ethylene Propylene
HDSL	High-speed digital subscriber line
IEEE	The Institute of Electrical and Electronics Engineers
ONT	Optical network terminals
PoE	Power over Ethernet
POTS	Plain old telephone service
PVC	Poly Vinyl Chloride
RFI	Radio frequency interference
RG	Radio guide
RJ-45	Registered Jack 45
STP	Shielded twisted pair
TAT-1	Transatlantic No. 1
UTP	Unshielded twisted pair
WDM	Wavelength division multiplexing

5.1 INTRODUCTION

Transmission medium is an essential component of a network along with a transmitter and a receiver without which the communication cannot be completed. The transmission medium is a component that serves the purpose of telecommunication by giving a path to data signals to pass through and transmit from one place to another. Therefore, the transmission medium can be said as the highways or arteries of telecommunication. Transmission media are either wire-lined or wireless; copper cables and optical fiber cables are popular types of the wireline medium. Both the copper cable and optical fiber cable can provide connectivity to homes and business with the Internet. But nowadays, optical fiber cables seem to be more and more popular and efficient in data centers requiring high data speed [1]. Optical fibers have several advantages and can provide greater bandwidth and longer transmission distances with low signal losses. The optic fiber cable is regarded as the best transmission medium since its introduction. This technology works by fiber cables made of glass or plastic strands which can transmit light. This technology allows the data or

DOI: 10.1201/9781003302902-5

voice signals to go through the cables as the beam of light rather than electrical signals. This can deliver more data and has good quality of signals or increased degree of integrity over longer distances as compared to copper cables. Optical fibers are immune to electrical interference or noise/disturbances, making them more reliable and safer. This is the reason that most of the connections between servers have been replaced with optical links in data centers.

However, with the development of technology like power over Ethernet, the copper cables also remain vital in some of the networking applications, e.g., Internet Protocol camera or in-building networks, considering that they can deliver data and supply power simultaneously.

So, it is a remarkable question to answer: which among them, the copper cable or the optical fiber cable, is the choice of future? Will the copper cables be replaced by optical fiber cables? To compare things is a general tendency of human and so it is common to compare among the transmission mediums and say that one transmission medium is better than another [2]. If we analyze, we will find that each of the media systems has some advantages as well as some flaws, and the best media system can be based on the purpose of communication system and the desirable end result. However, each of the transmission media has played a role in the evolution of telecommunication and has a significant place in the history itself. At the same time, each has some of those characteristics that will make it the ideal medium to use in certain circumstances. Cost, ease of installation, operations, maintenance, availability, noise immunity, and efficiency of transmission are some of the important factors while considering a suitable transmission medium.

5.2 HISTORY OF TELECOMMUNICATION CABLES

Considering the history of telecommunication, it can be said that the modern-day communication has been started with the invention of wire-lined telegraph by Samuel Morse in 1832. The technology developed by Morse was able to transfer the information instantly over long distances. The telephone was invented by Alexander Graham Bell over 100 years ago. Bell invented the telephone in 1876, and later, he took a patent on copper twisted pairs in 1881, and this was the birth of the modern telephone system [3,4] (Figure 5.1).

During the early days, the telephone lines were used to connect pairs of telephones through separate lines such that each person having a telephone can talk to one another through it. Historically, the first telephone line was set up in Boston in 1877, which was established between the Boston office of Charles Williams Jr. and his home in Somerville, MA [5].

However, this way of setting up the private lines for telephone connection was, of course, very restrictive and expensive that everyone could not afford, and the people were willing to be able to talk with more than one household via this one-to-one telephone connection. This was of course needing a type of central interconnection facility with the push from the government and through the participation of certain stakeholders, which resulted in the concept of public telephony. This was to allow the capability of connecting the telephone lines as and when desired, which resulted in the concept of circuit switching in telecommunications. Furthermore, the telephone

FIGURE 5.1 Alexander Graham Bell's first telephone, 1876.

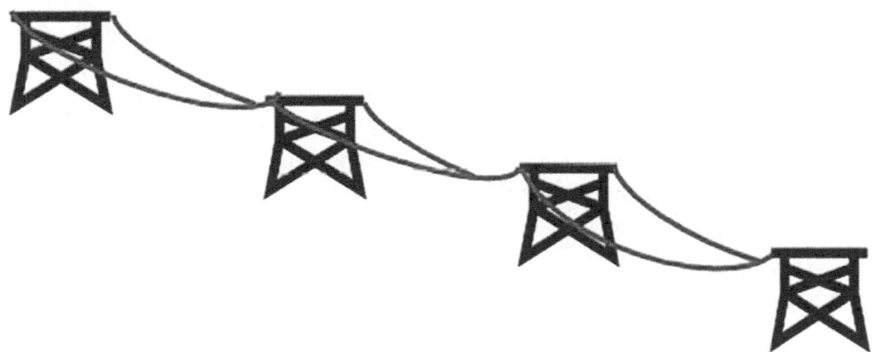

FIGURE 5.2 Diagram representing typical suspension cable.

lines were supposed to connect the central interconnection facility with the private telephones owned by the end users.

Figure 5.2 represents a typical suspension cable for transmission over the poles. The telephone lines were initially the overhead lines which were placed on telephone poles or attached to racks on rooftops and made of single-grounded iron or steel wires. The initial telephone lines were inherently noisy single-grounded wires. Further, attempts were made to decrease the nose in the telecommunication lines by making the lines with phosphor bronze wires and compound copper steel. Although the benefits of the copper wire were known, the technology to make an overhead transmission copper wire was not available to make a wire strong enough for suspension over the poles. Thomas Doolittle developed a technique of hardening the copper wire in 1877, by annealing copper wire to increase its tensile strength drawn through a series of dies. The copper wire made using this technique was strong enough as the suspension wires and copper wires became dominant in the telephone market. Using the copper suspension wire, a long-distance experimental telephone line was set up between Boston and New York in 1884; this was the beginning of copper telephone lines. Further copper telephone line connection was established between New York and Philadelphia in 1885, and many of the cable manufacturers started mass manufacturing of the copper telephone cables. With its application in power transmission, this hard-drawn copper cable has also expanded the electrical industry [5].

Since the invention of "optical telegraph" by Chappe brothers in 1790, the concept of optical communication has been available. This was developed with towers fitted using a series of lights to relay messages back and forth. This technique may not be comparable to today's sophisticated technique of optical communications using fibers but was one of the very earliest steps toward optical communication using light as a signal carrier. Many of the physicists and scientists further contributed to conceptualizing, developing, and standardizing the technique of optical communication in the history. A significant step for the development of modern fiber cabling happened when Abraham Van Heel produced a cladded fiber system in 1954 that greatly reduced signal interference between fibers. However, the first successful version of modern fiber-optic cable development did not occur until 1970. The nonexperimental version of fiber-optic cables was developed to be deployed by telephone companies for rebuilding the communications infrastructures by the 1970s and 1980s. Furthermore, the advancement in the bandwidth and distance capabilities made the fibers significantly less expensive than any other communication mediums, due to which the replacement work of traditional cables with fibers started by the mid-1980s. Cable television also discovered the opportunity to enhance the performance and reliability using fiber cables in the mid-1990s and could also enable the offering of both phone and Internet services through the same fiber cable.

5.2.1 POTS

Plain old telephone service (POTS) is an old system of communication, representing a voice-grade telephone service using analog signal transmission over copper loops. POTS was initially the acronym for "post office telephone service", as earlier, the post office was the conventional mode of communication from one place to another. Also, initially during those days, callers relied on post office operators, who were supposed to connect the phone lines to intended destinations. Later, that service was taken away from the national post offices, and the term "plain old telephone service" was then adopted from "post office telephone service".

The transmission of voice-grade analog signals takes place over a pair of twisted copper wires in POTS [6]. It is the phone line technology with copper wires carrying our voice that most of us grew up with. We can still remember POTS in our old telephone system or landline phones where a crammed combination of thick wires dangling overhead, hanging between poles was the POTS. That crammed combination of thick wires was the copper wires through which the voice signals traveled from

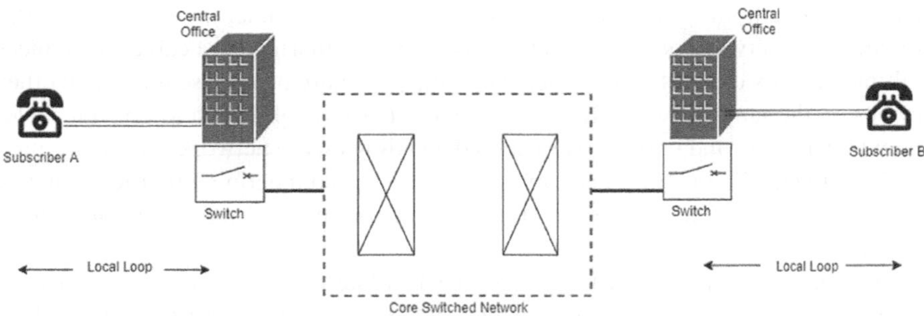

FIGURE 5.3 Diagram representing a typical POTS.

one place to another. The network of POTS traversed from countries and continents over copper cables to facilitate voice communication. It was the standard voice-grade telephone system used by residences and businesses across the world since its inception in the 1880s.

Figure 5.3 represents a typical POTS system.

To transmit the voice through a POTS line, the sound waves are to be converted into electrical signals first, and the electrical signal is to pass through the network. Copper wires, as the twisted pairs, are used to transmit analog electrical signals. This also needs a dedicated switch to establish the line connection based on circuit switching for the signal to travel through. The dedicated circuit connection is reliable, whereas the established line is reserved only on per call basis. It also requires the operators to plug wires into a common patch panel to connect two parties while forming a circuit switching connection. A trunk connection is one that requires the operators of two exchanges to plug the caller and receiver wires circuit connection into the same the line at the same time. Then came the automated switching that worked by responding to signals from a calling device, eliminating the need of operators for circuit switched connections.

These POTS system is an upgrade over the rudimentary first edition phone system invented by Alexander Graham Bell. We can find the communication between two parties was originally dependent on an operator to manually connect the call. However, the switching aspect has been changed with time and technology evolution. This is now being automated, and the system is almost entirely digital.

5.2.2 Development of Ethernet

We discussed about the Ethernet technology in detail in our last chapter (Chapter 4). Robert Metcalfe at Xerox Palo Alto Research Center, after getting inspired from ALOHAnet, in the 1970s started working on the cabled version of the Ethernet protocol and in 1976, presented an experimental Ethernet. Ethernet is the remarkable achievement in the history of networking system that enables to connect a number of computers and other devices over a single network, providing a high-speed connection. Thus, Ethernet has a very important role in making easier communication between devices and is a very popular global standard comprising a system of wires and cables to adjoining multiple computers, devices, and machines over single the organizations network or local area network (LAN). Ethernet uses cables to connect multiple network devices, two or more computers to share printer, scanner, and other devices on the same network. It uses different types of cables such as coaxial cable, optical fiber cable, and twisted-pair cables depending on the network's topology, protocol, and size. These Ethernet cables are used for connecting Ethernet switches and Ethernet routers to network devices, computers, servers, and so on through an Ethernet interface.

To standardize the Ethernet protocol, the Institute of Electrical and Electronics Engineers started a project working group 802.3 specifically for Ethernet development. Before the Ethernet standards (802.3 specifications approval from the Institute of Electrical and Electronics Engineers project committee) getting commercialized, many of the LAN innovations such as Cheapernet, Ethernet-on-Broadband, and StarLAN arrived that started gaining attention and popularity from many manufacturers between 1983 and 1986. This resulted in use of Ethernet networks in the workplace by some small companies, which was still via telephone-based four-wire lines. Ethernet cables used to connect offices and homes typically were twisted-pair cables such as Cat 5, Cat 6, and Cat 7. Twisting of the wires together has a significant effect that enables the currents to balance and enable the overall fields around the twisted pair to cancel.

Ethernet networking systems have ruled the networking industries and have a rising market globally adapting to the latest technology innovations. The growth of the global Ethernet cable market is driven by the key factors such as the advantage of higher speed and lower latency, improved reliability and low security threats, and easy installation and connection. A significant preference was given to the wired over wireless connectivity for high-speed data transfer to propel the global Ethernet cable market growth. That also offers improved reliability of these Ethernet cables to transmit data seamlessly. At the same time, a rise in demand is expected from the telecom industry for power over Ethernet cables that enable to carry electrical power along with providing data connection.

5.2.3 From Analog to Digital

The initial telephone lines and the voice-grade signals were totally analog signals and were carried through the electrical cables (copper wires) in the form of electrical signals. The continuous variation in signal amplitude and frequency characterizes an analog signal. When we speak into a handset, the air pressure around our mouth changes with our voice in the case of telephony. Those changes in the air pressure fall onto the microphone of the handset, which then amplify and get converted into an electrical signal as current or voltage fluctuations. This signal conversion from voice to current fluctuations is analog in characteristic and represents the actual voice pattern. The speech frequency range of the vast majority of intelligible sounds falls between 250 and 3,400 Hz, and therefore, a total bandwidth of 4,000 Hz was typically allotted for voice transmission by the telephone companies. However, the twisted-pair wire has a total capacity of 1 MHz of the frequency spectrum. Due to this reason, there was a requirement of putting a bandwidth-limiting filter on the analog circuit to filter out all frequencies above 4,000 Hz to provision a voice-grade analog circuit. Because of that, the analog circuits can conduct only low-speed data communications, and the maximum data rate over an analog facility is 33.6 Kbps [7].

There was no international standard for the telephone cable prior to the digital communications and Ethernet gaining widespread popularity and acceptance, and the standards were generally set up at the national level rather than globally, e.g., the UK had the cable specifications CW1293 and CW1308 for the General Post Office standard. Both types of the cable were of the similar standard as that of the twisted-pair Category 3 cable. The insulation of the telephone twisted-pair cable was done using wax-coated paper applied to the copper prior to the common use of polyethylene and other plastics for insulation with the overall lead sheath of this type of cable. Shortly after the invention of the telephone, such type of cable came into existence in the late 19th century. The cable termination was sealed using molten wax or a resin to prevent the ingress of moisture in termination boxes. Otherwise, the moisture would seriously degrade the insulating properties of the wax-coated paper insulation of the copper twisted-pair cable in the box. However, the future maintenance and changes of such type of cable with that sealing and insulation were more difficult and are currently no longer made.

Majority of the long-distance communication was provided by analog systems only using frequency-division multiplexing until the early 1980s. However, the development of fiber optics has a very important role in rapidly replacing those analog systems with digital systems, whereas the digital transmission is also carried over the coaxial and microwave systems, where the telephone signals are first converted to digital (quantized and discrete time format) from an analog signal. Digital signal transmission is different from analog one; it has a series of discrete pulses that represent bits as 1 and 0 rather than being a continuously variable wave form. The Bit 1 is represented as high voltage and the Bit 0, as null or low voltage in electrical networks. Whereas Bit 1 is represented by the presence of light and Bit 0 is represented by the absence of light in the optical networks. It is comparatively easier to reproduce the digital signal than an analog signal, and the signal transmission is simpler. Digital transmission needed the signal regenerators rather than dumb amplifiers of analog transmission. The T-1 line of the digital transmission is characterized to carry data at a rate of 1.544 Mbps, and an E-1 line can transport data at a rate of 2.048 Mbps.

5.2.4 Transatlantic Cables

Transatlantic telegraph cable installation was done using the undersea cables under the Atlantic Ocean for telegraph communications that used the coaxial cable for transatlantic cable installations in the first time in 1858. Telegraphy became obsolete communication, but telephone and data are still carried on other transatlantic telecommunications cables, where the first cable was laid across the Atlantic from Valentia in western Ireland to Bay of Bulls, Trinity Bay Newfoundland in the 1850s. The Atlantic Telegraph Company and Cyrus West Field were involved in constructing the first transatlantic telegraph cable, the project which began in 1854 and completed in 1858; the transatlantic cable route is shown in the map in Figure 5.4. Queen Victoria and the US president, James Buchanan, exchanged telegraphic pleasantries to inaugurate the first transatlantic telegraph cable on August 16, 1858, where queen's

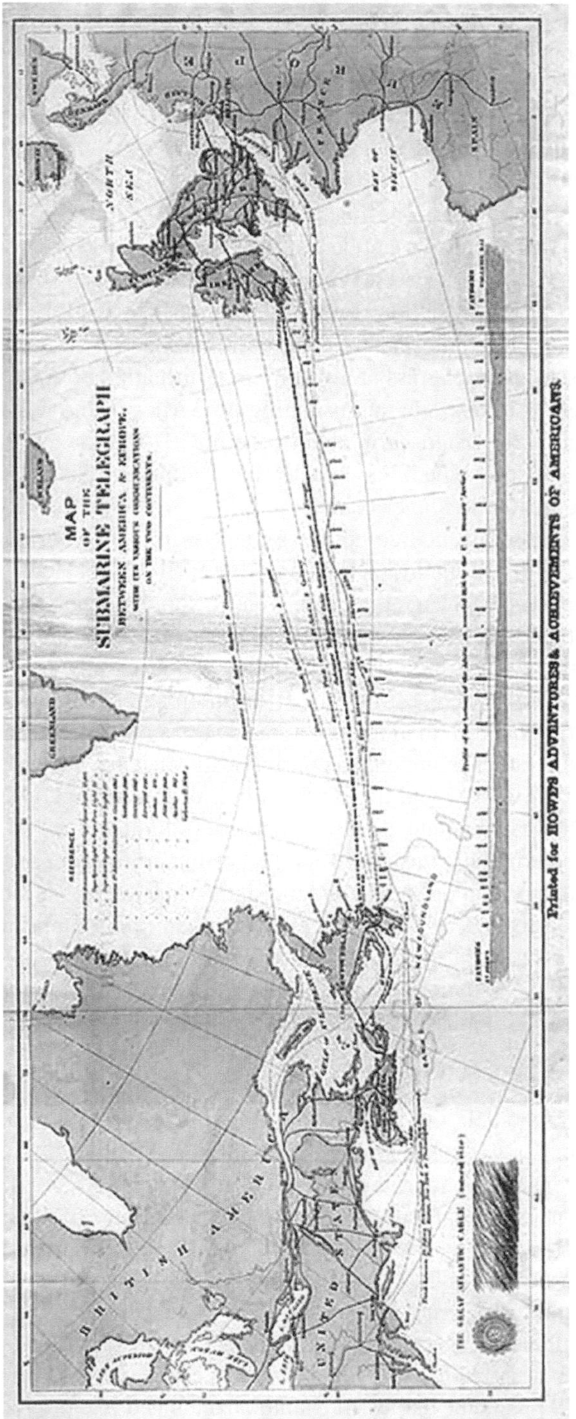

FIGURE 5.4 Map of Transatlantic cable route in 1858. (Wikipedia.)

98-word goodwill greeting took almost 16 hours to send through the 3,200 km long cable. Unfortunately, the cable could only function for 3 weeks because the cable's insulation failed, and it had to be abandoned [8].

The second attempt was made for the transatlantic cable with much of the material improvement in 1865, where the cable was laid with the ship SS Great Eastern. Unfortunately, this cable broke more than halfway across, and they had to give up after many rescue attempts. A third attempt for transatlantic cable was again laid by the Anglo-American Cable house in July 1866, and the successful connection was put into service on July 27, 1866. Furthermore, the 1865 cable was also retrieved and spliced, resulting in two cables that were more durable and had good line speed.

The first transatlantic telephone cable system was Transatlantic No. (TAT)-1, laid by the cable ship Monarch between Gallanach Bay, near Oban, and Clarenville, Newfoundland during 1955-1956. This transatlantic telephone cable system was made of using two coaxial cables, one for each direction, and used analog frequency-division multiplexing to carry 36 telephone channels two-way voice circuits. It was inaugurated on September 25, 1956, and had 588 calls made from London to the US, whereas 119 calls were made from London to Canada in the first 24 hours of its operation [9,10]. Transatlantic telephone experienced a dramatic traffic increase with the availability of the cable system from 1.7 million calls in 1955 to 3.7 million in 1960. Six additional coaxial cables were laid across the Atlantic Ocean between 1956 and 1983 that were representing four successive generations of cable design.

The cable industry has seen major technology shifts in the undersea telecommunications which allowed tremendous capacity growth since the first telephone cable introduction in the 1950s. One major technology shift was from the analog coaxial cable to digital single-mode fiber systems in the 1980s. Since this change was an important technology challenge, the capacity increase was modest initially. It was well understood by that time that the fiber-optic transmission systems had great potential capacity to be explored. However, it had the "bottleneck" of using the electro-optic regenerators that prevented realizing the full potential of fiber-optic systems, and later, the fiber-optic undersea cable had the revolutionary technology shift with the introduction of the optical amplifier–based repeater in the 1990s. Fiber-optic undersea cables had seen tremendous magnitude of capacity growth starting with the ability to operate a single channel at ten times the rate of the regenerated systems using wavelength division multiplexing with wide-band optical amplifier repeaters and continues today with coherent transponder technology. The first optical amplifier–based transatlantic system TAT-12/13 with a single 5 Gbps optical channel per fiber pair was commissioned in 1996. One of the more ambitious programs of the transatlantic undersea fiber-optic cable was the TAT-14, deployed in 2001. It had connected the US, France, Germany, Denmark, and the UK with a 15,428 km (9,581-mile) undersea cable, having four fiber pairs with a protected capacity of 640 Gbps, roughly 9.6 million voice circuits.

5.3 COPPER CABLES

The electrical properties of the copper medium create resistance and interference for the electrical signal passing through it; this resistance slows down the signal or flow of current. Due to the electrical properties of the copper medium, the further communication signals travel, the more they are weakened. Thus, in terms of limiting the transmission speed and the distance of the copper cable, electrical properties of copper are key factors. On the other hand, it is the same electrical properties of the copper medium along with costs, availability, manufacturing ease, the ability to make a thin strand, and so on. That made copper as a logical choice of communication media [11].

In the electrical wiring, the use of copper started with the invention of the electromagnet as well as the telegraph in the 1820s. The demand for the copper conductor in the electrical wiring further increased with the invention of the telephone in 1876 as an electrical conductor. Copper cables have an extensive and vast range of application as the electrical conductor in many categories of electrical wiring such as generation of power, transmission of electricity, power distribution, telecommunications, electronics circuitry, and so many countless electrical equipment. One of the most important markets of the copper industry is the electrical wiring in buildings.

For the wired connectivity of home since the telephone's advent over 100 years ago, the dominant technique involved was based on using the copper cabling. The copper cable used in telephone was considered as perfectly suitable for a voice signal for which it was intended to be the telephone lines transmission medium. However, it has a limited bandwidth provision considering all the other things. The world was so much familiar with copper cables for connectivity that many people doubted any other medium would ever be able to replace copper cables for connectivity until fiber optics came along.

Some of the interesting facts of copper in communications are as follows:

Fact 1: It was thought not long ago that only optical fiber cables could handle the big bandwidth requirements, and copper has to be replaced wherever there is a high-bandwidth need. But with the latest innovations in copper technology, communication between computers can now achieve data speeds up to 10 Gbps on twisted pairs of copper wire called structured wiring.

Fact 2: The development of the high-speed digital subscriber line– and asymmetrical digital subscriber line–based technologies has enabled the telephone companies to capitalize on existing copper lines. This in turn enabled businesses to accommodate with the lower-cost copper-based networking options without needing to switch to high-cost fiber optics. Hence, this technology also allowed the voice and data transmissions to be conducted simultaneously on the existing copper phone wires.

Fact 3: The structured wiring of Category 6 or better variants allows maximum advantage of computer-based technologies to the users. Four pairs tightly twisted copper cable of #24 gauge insulated copper conductors is the most common jacketed cable that is extensively used in commercial applications and can achieve greater capacities of data speed up to 10 Gbps.

5.3.1 Twisted Pairs

The twisted-pair cable is one of the very older and popular copper cables used since the history of copper cables. It is a type of copper pair cable which is made by running two separate insulated copper wires parallel to each other in a twisted pattern. That cables are

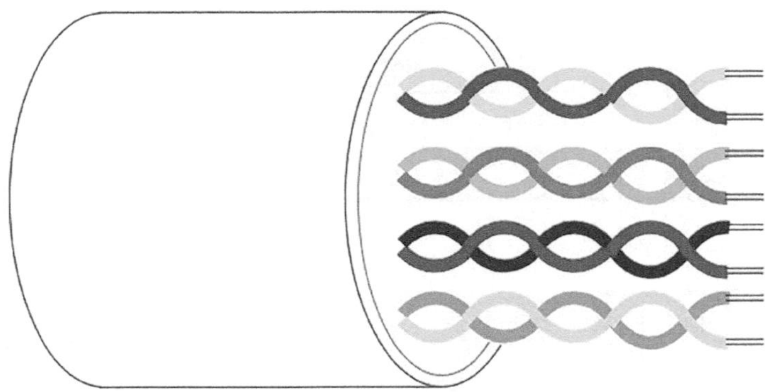

FIGURE 5.5 Twisted-pair cable.

formed by twisting a pair of copper wires together, and the pair of wires forms a circuit to transmit data. That pair of cable after twisting has a significance in terms of the electrical signal carrying capacity. This type of cabling fulfilled the need for telephone communications and is still used in the majority of the communication networks. It is widely used in different kinds of data and voice communication network infrastructures (Figure 5.5).

With twisting of the pair of copper wires, it has the advantage of providing protection against crosstalk – the noise generated due to adjacent copper pairs. Copper is a good electrical conductor, and we are using its electrical properties to transmit the communication signals. With the passage of electrical current through a copper wire, a small circular magnetic field is created around the wire. When we twist the two copper wires placed close together in an electrical circuit, the magnetic fields associated with each of the wires twist are exactly opposite to each other. With that, the twisted copper pair gets the capability to cancel each other's magnetic field, and it can also cancel the effect of any outside magnetic fields, hence providing noise immunity to the actual communication signal passing through the cable. Cable designers use this technique of cancellation of noise with twisting the wires to effectively provide the self-shielding capability of the copper wire pairs within the network media.

This cabling is often used in such data networks that require short- and medium-length connections. The twisted-pair cables are cost effective in telecom applications, i.e., they offer relatively lower costs than the optical fiber and coaxial cable (Figure 5.6).

There are two types of twisted-pair cables: unshielded twisted pair (UTP) and shielded twisted pair (STP). We will discuss on the UTP and STP cable in more detail below.

UTP: This type of cable can be found in many Ethernet networks and telephone systems with having a vast range of applications. We discussed about the twisted pairs of copper cables earlier; the unshielded twisted pair is the basic type of twisted pair that uses the twist of the copper wire pair to control the electrical interference and noise. The UTP cable has no additional shielding provided to the twisted copper pairs, so it is known as the UTP.

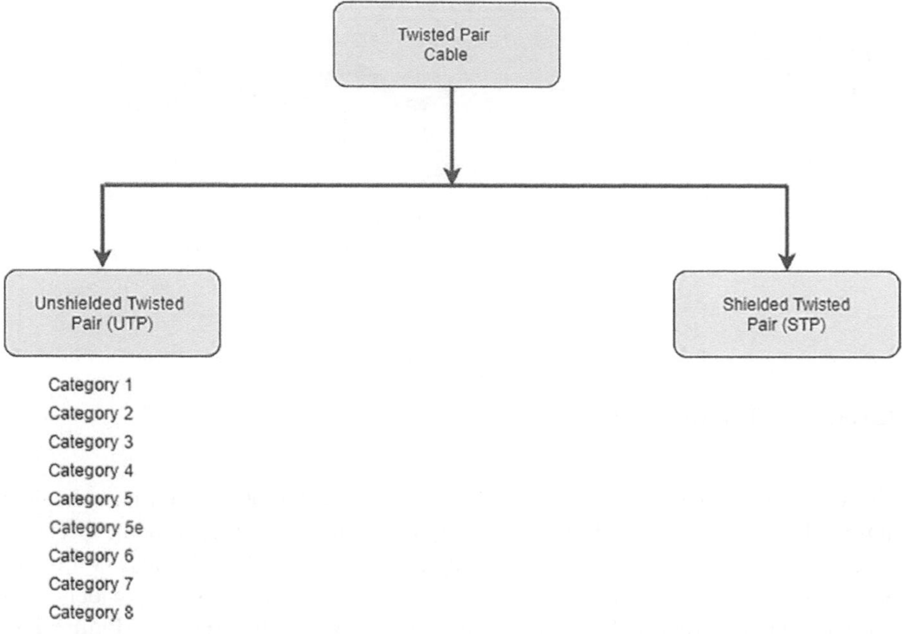

FIGURE 5.6 Twisted-pair cable types.

Therefore, to limit signal degradation caused by electromagnetic interference and radio frequency interference, the UTP cable has to solely rely on the noise cancellation effect due to the twisted wire pairs. The number of twists in the copper wire pairs varies to reduce further crosstalk between the pairs of the UTP cable. While designing the UTP cable, the cable designers have to follow precise specifications that govern how many twists or braids should be permitted on per meter of the cable.

It is also the most commonly used cable in computer networks; the data networking standard of modern Ethernet can use UTP cables. With the bandwidth improvement of the UTP cable to match the baseband of television signals, it is also being used in some video applications like CCTV, security cameras, and so on (Figure 5.7).

The UTP has also been grouped into a set of 25 pairs for indoor telephone applications. This grouping was in accordance with a standard 25-pair color code which was originally developed by AT&T Corporation. Out of these color code–based grouping, a typical subset of color code combination, e.g., white/blue, blue/white, white/orange, and orange/white, is often shown in many UTP cables. These UTP cables are using the copper wire as the central conductor to work as the signal transmission medium, measured at 22 or 24 American wire gauge. The copper wire is further coated with a colored insulation layer; that insulation layer is typically made from polyethylene or FEP insulators. After that, the entire package, including all the insulated and twisted copper wires, is further covered in a polyethylene jacket or PVC

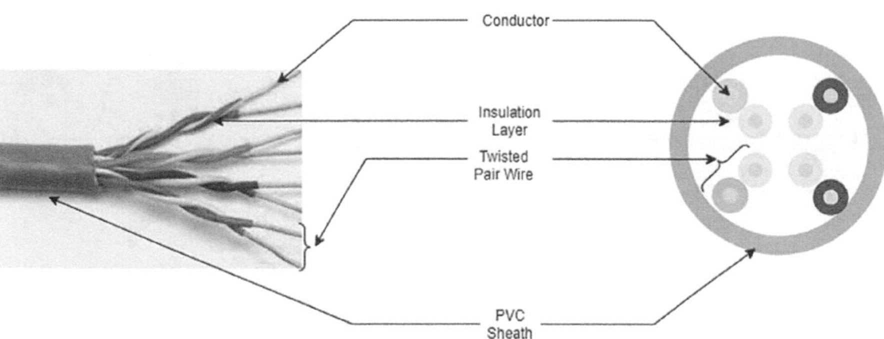

FIGURE 5.7 Unshielded twisted-pair cable.

sheath. The cable is divided into identical bundles for urban outdoor telephone cables, which contain hundreds or thousands of pairs. Each bundle contains twisted pairs with different twist rates, due to the reason that the pairs with the same twist rate within the cable can still get some degree of crosstalk. And these bundles are further twisted together to make up the cable.

Registered Jack 45 (RJ-45) connector is often used to install the UTP cable. The RJ-45 connector is an eight-wire connector that is used very commonly to connect computers onto a LAN or Ethernets.

There are several categories of the UTP cables that have been developed, which range from Category 1 to Category 8 UTP cables. These categories of the UTP cable are based on the different number of copper wire pairs used, difference in the data transmission rate, and difference in the implementation. With the increasing data rates of the modern day's application, the requirement is fulfilled by higher specification variants of the UTP cable. Categories 5, 6, 7, and 8 have become very popular for some new emerging applications in recent years. Category 5 is one such popular cable among them, which offers widely supported operations and provides reasonable price.

STP: This type of cable is composed of having a fine wire mesh surrounding the insulated twisted-pair copper wires to protect the transmission with an additive layer of shielding provided. Due to this reason, this type of twisted-pair cables is called as the STP which the UTP cable does not have. It is used in older telephone networks and data communications for the purpose of reducing the outside interference (Figure 5.8).

Shielding of the twisted pairs provides an electrically conductive barrier that helps in attenuating the shield's external electromagnetic waves. In most occasions, shielding of the twisted pair also functions as a ground, such that a conduction path is provided by the shield to induce current that circulates and returns to the source from that ground reference connection. Such type of shielding of the twisted copper pairs can be applied to individual twisted pairs or to a collection of twisted pairs. Shielding of the twisted cable pairs is done by using a metallic substance in an STP cable. And sometimes, all the pairs (typically four twisted pairs or eight strand cable) are then wrapped

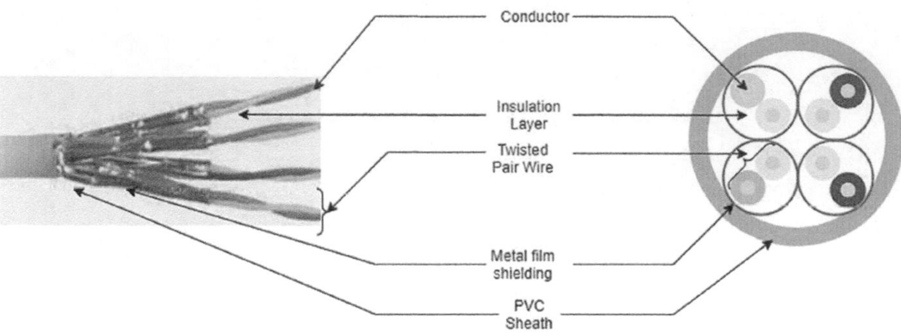

FIGURE 5.8 Shielded twisted-pair cable.

TABLE 5.1
Comparison between STP and UTP

Criteria	STP	UTP
Definition	STP cable is a twisted pair enclosed in foil or mesh shield	UTP cable is with wires twisted together
Crosstalk	Lesser due to noise and interference cancellation effect from the shield	Higher as compared to that of STP
Grounding	Proper grounding is necessarily required	Not required
Installation	Difficult installation as compared to that of the UTP	Easy cable installation
Cost	Expensive as compared to UTP	Cheaper than STP
Maintenance	High	Less compared to STP
Data Rates	Faster, high data rates	Slower than STP
Usage	Used for connections among enterprises over longer distance for data transmission	Used for transmission of data within short distance, home networking

again in another metallic protector. We can observe that, in general, there are three different techniques applied with the STP cable: shielding, cancellation, and wire twisting to prevent interference. This helps in protecting the STP cable from crosstalk and also increases the cable's fidelity.

The interference prevention ability of the STP cable is better than that of the UTP cables, and at the same time, the STP is more expensive and difficult to install. Also, STP has the need of grounding metallic shield at both ends of the STP cable. Otherwise, the shield may behave like an antenna and can pick up unwanted signals if it is not grounded properly. Due to the reason of being expansive as compared to UTP and the difficulty associated with their termination, STP cables are not so commonly used in Ethernet networks. Most of the common use case of STP cables are for the connection between equipment, racks, and buildings.

Comparing STP cables with UTP cables (Table 5.1).

5.3.2 Coaxial Cables

A coaxial cable is another type of cable which is popular in telecommunication and data networking. This type of cable has an inner conductor in the core of the cable surrounded by a layer of insulating material. The insulating layer is yet surrounded by another layer of a conductor foil to provide conductive shielding. The entire cable is then packaged in an insulating outer PVC jacket or sheath. It is referred by the term "coaxial" which describes about the inner conductor and the outer shield are sharing the same geometric axis and hence are coaxial to each other. Consider Figure 5.9 which represents a typical coaxial cable with its cross-sectional view. Data transmission happens with the electrical signal passing through the inner conductor present in the core of the cable.

The coaxial cables have been around us from a long time as a technology (since the early 20th century) and provide many advantages for reliable and low-loss signal transmission. As the transmission line coaxial cable is used to carry low losses and high-frequency electrical signals, it has its usage across wide range of applications such as telephone trunk lines, broadband Internet cables, computer data busses with high speed, cable television, and connecting radio transmitters and receivers through antennas. Therefore, the cable operators, telephone companies, and Internet service providers around the world commonly use the coaxial cable to convey data, video, and voice communications. At the same time, the coaxial cable is also used extensively within homes.

The coaxial cable differs from other available shielded cables due to the reason that the dimensions of the cable and connectors can be controlled to give precise and constant conductor spacing to help the cable to function efficiently as a transmission line. The success of the coaxial cable is due to the shielded design that allows the copper core conductor to transmit data quickly without interference losses or damage in the signal from environmental factors. An important feature of the coaxial cable is its great capacity to carry many channels simultaneously, without needing thousands of separate wires. This was the reason that the cable television companies found its great use in bringing many channels of subscription television to the residences of the subscribers.

The coaxial cable has some limitations as well which cause it to be replaced with the optical fiber cable in some cases or with category cables. One disadvantage of the

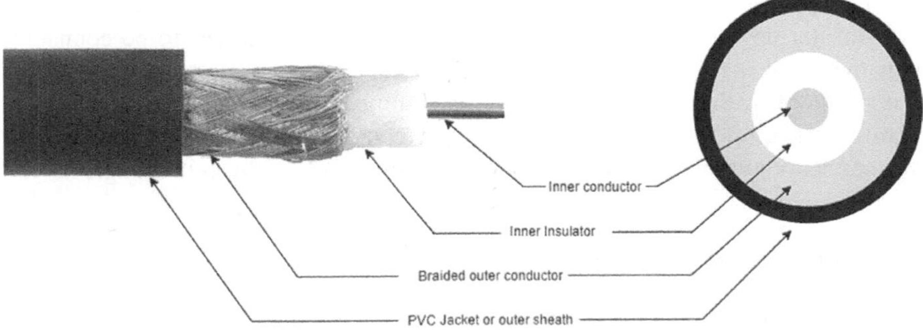

FIGURE 5.9 Coaxial cable.

coaxial cable is that the cable must be screwed onto the electronic unit because of which installation of the cable can sometimes be difficult.

Radio guide (RG)-6, RG-11, and RG-59 are the three most common cable sizes of the coaxial cable. These numbers represent the diameter in the versions of the RG cable. Impedance is an important characteristic of the coaxial cable, and most of coaxial cable specifications have an impedance of 50, 52, 75, or 93 ohms. RG-6 cables having impedance of 75 ohms and with double or quad shields have become a de facto standard for many industries.

5.3.3 Hybrid Cables

Hybrid cables are the innovative developments of transmission cable technology that combined the feature of fiber optic for high-speed data requirement and power transmission requirement through copper cables. These hybrid fiber-copper–based solutions have opened new possibilities of cable for transmitting both data and power through a single cable [12,13]. Although we already had the traditional method as an option to run fiber cable for lossless long-distance high-speed data transmission and then use of a separate local source for power, there are certain benefits of using the hybrid cable instead. Some of the advantages are hybrid technology requires only a single cable to pull, manage installation, and offer labor savings due to data and power are combined into one cable. No need of paying to run low-voltage power at each location as hybrid cable itself brings the power to these locations without any requirement of a local power source, and the power is instead to be carried via the cable from a centrally managed remote source. A hybrid cable can be used to carry power and data to the pole-mounted external security camera on the light pole with a single hybrid cable instead of two, without worrying about performance and distance limits.

Hybrid fiber-copper cables are available with a variety of design techniques that include the multimode or single-mode fiber. Depending on the application and number of devices to be connected, the hybrid cables can include a single fiber or multiple fibers. Consider the example of optical network terminals which are used in passive optical networks operating over a single fiber. The signals in the optical network terminal are transmitted simultaneously in both directions over separate wavelengths based on wavelength division multiplexing with 1,310 nm for upstream data and 1,490 nm for downstream data. Depending on how many devices to be connected with their power requirements, the copper conductors also range in number and type for power within hybrid cables. Consider some hybrid cables that have a power requirement of up to 12 copper conductors to connect a remote power supply unit and another requirement of only two copper conductors to connect to a single device.

5.4 OPTICAL FIBER CABLES

An optical fiber or fiber-optic cable refers to that technology of communication which transmits data through thin strands of a highly transparent material, which is usually either glass or plastic. It was launched in the 1970s, but the first optical fiber telecommunications network was not installed until the early 1980s. The optical fiber cable is a high-speed data transmission medium of modern age communication that contains tiny glass elements within the cable coated with plastic layers to carry light beams. Its assembly is similar to that of an electrical cable, but in place of the metallic conductor, it contains one or more optical fibers that are used to the carry the signal in the form of light waves rather than electrical pulse in the electrical cable. The fiber elements of the optical fiber cable are individually coated with plastic layers that are further placed in a protective tube suitable for the transmission of light waves for data transmission.

It is a very thin strand of pure glass to act as a waveguide for light waves and can carry them over long distances. The transmission of light waves through the optical fiber is based on the characteristic of total internal reflection of light. The optical fiber cable is composed of two layers of glass and consists of a core and a cladding layer to work as a light waveguide. The "core" glass layer of the optical fiber cable carries the actual signals as light waves, and the "cladding" is the surrounding glass layer of the core. While designing the fiber, the refractive index is selected such that the difference in the refractive index between the two mediums can allow total internal reflection of light waves when passed through. The coating of cladding is usually done with a layer of an acrylate polymer or polyimide in practical fibers. This is done to protect the fiber from damage and does not contribute to the optical waveguide characteristic of the optical fiber. A cable is formed with adding the several layers of protective sheathing, depending on the application of the optical fiber cable. Sometimes, forming of the fiber cable is done with putting the light-absorbing dark glass or rigid fiber assemblies between the fibers. This helps in preventing the leakage of light from one fiber to another or reducing the crosstalk between the fibers (Figure 5.10).

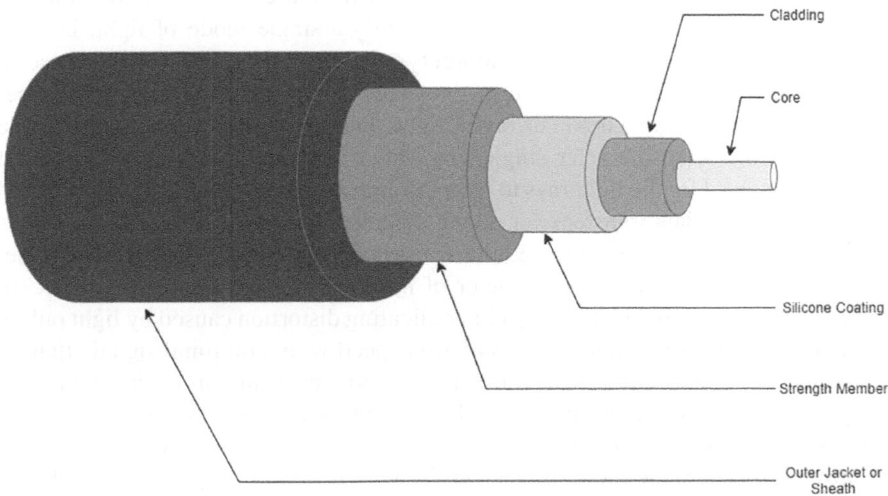

FIGURE 5.10 Typical structure of an optical fiber cable.

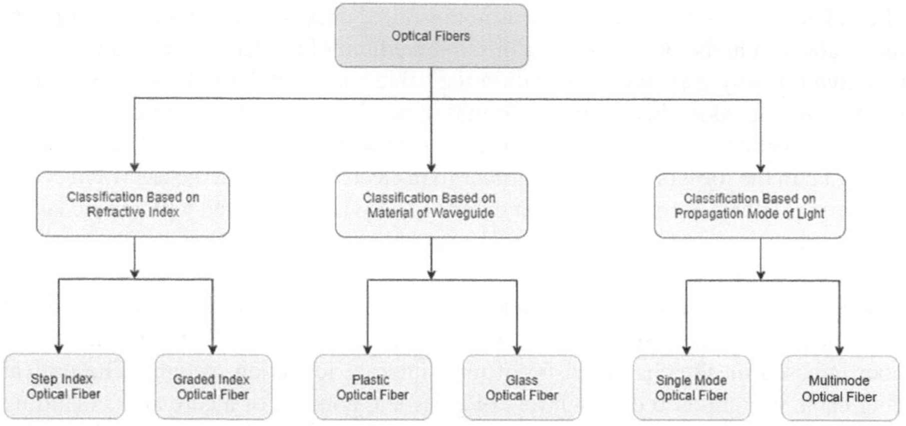

FIGURE 5.11 Classification of optical fiber cable.

An optical fiber cable provides the transmission of digital data through the cable as rapid light pulses. Optical fiber cables are used in a variety of ways; for different applications, different types of the optical fiber cable are used. For example, the different optical cable has to be used in long-distance telecommunication and provide a high-speed data connection between different parts of a building. Submarine cables or dark fibers are one of the popular use cases of long-distance or intercontinental wired communication, considering the example of the transatlantic telecommunications cables between the US and Europe which are mostly with the optical fiber [14] (Figure 5.11).

5.4.1 Single-Mode Fiber

A single-mode optical fiber is a type of optical fiber used in fiber-optic communication that is designed for transmission of only a single mode of light, i.e., the transverse mode, through it. It is a common type of the optical fiber which is used to transmit the optical signals over longer distances, composed of a single glass fiber strand to transmit a single mode or ray of light. The single-mode optical fiber cable has a fairly slim core diameter single glass fiber strand. It has very less internal reflection involved for the light rays to pass through it, which reduces attenuation and allows high-speed data transfer over a long distance. It is also sometimes known as a unimode optical fiber, monomode optical fiber, or single-mode optical waveguide. A single-mode fiber has a core diameter of nominally 9 μm, where its small core coupled with a single light wave helps in eradicating distortion caused by light pulses overlapping. It offers the highest transmission speed with a minimal signal attenuation, where the modes of the light waves are the possible solutions of the Helmholtz equation obtained by combining Maxwell's equations and the boundary conditions, which further define the distribution of the wave in space or the way the wave travels through space. Same mode waves can be with different frequencies which can be used to propagate in the single-mode fibers. Waves having the same mode and different frequencies are similarly distributed in space, which means that the same mode light with all the different frequency waves will give a single ray of light. The single-mode fiber is also called as the transverse mode due to the reason that the electromagnetic oscillations of the waves occur perpendicular (transverse) to the length of the fiber. Charles K. Kao was awarded the Nobel Prize in Physics for his theoretical work on the single-mode optical fiber in 2009 [15].

5.4.2 Multimode Fiber

A multimode optical fiber is that type of the optical fiber which is mostly used for short distance communication such as within a building or a campus. The multimode optical fiber is generally used for backbone applications in buildings due to its high capacity and reliability. It can be used to provide higher data rates up to 100 Gbps to the users across moderate distances since the single-mode fiber is more expensive than multimode fiber and is not suitable for short distance communication. It has a much larger core diameter which allows the propagation of the multiple light mode, and this is the main difference between multimode and single-mode fibers. The large core diameter also limits the maximum length of a transmission link because of modal dispersion. The core diameter of the multimode fiber typically ranges from 50 to 100 μm that is much larger than the wavelength of the light carried in it.

5.5 CONCLUSION

The transmission medium is a very essential and integral part of the evolution of communication technology, and improving the cable technology has effectively improved communication technology. We have gone through several of the communication media as presented in this chapter along with their use cases in telecommunication. From setting up of the telephone network in the 19th century, copper cables have played a very crucial role in providing a suitable transmission medium. The early telephone networks had limited subscribers, and the cables were generally single-line copper cables with limited capacity. The next step in network development was during the interwar period that had installation of long-distance cables for connecting European regions and nations [16]. Further technical innovations in the telephone networks resulted in the creation of wide-ranging telephone networks with automatic switching, duplex cables, increased efficiency, reduced costs, and enabled more and more people to use telephone networks. Various telephone lines of twisted-pair copper cables in the lower regional and local networks and coaxial cables in the upper regional and long-distance networks were used in the 1950s and 1960s. Copper cables have been the dominant technique since the telephone's advent over 100 years ago.

The coaxial cable is another type of cable which is popular in telecommunication and data networking. This type of cable has an inner conductor in the core of the cable surrounded by a layer of insulating material.

The optical fiber cable provides a high-speed data transmission medium of modern communication that contains a tiny glass element within the cable coated with plastic layers to carry light beams. The optical fiber cable provides the transmission of digital data through the cable as rapid light pulses.

Hybrid cables are the innovative developments of transmission cable technology that combined the feature of the fiber optic for high-speed data requirement and power transmission requirement through copper cables. These hybrid fiber-copper–based solutions have opened new possibilities of cable for transmitting both data and power through a single cable.

REFERENCES

1. The history of fiber optics in telecommunications, May 10, 2018. Carritech Telecommunications. Accessed: Apr. 6, 2021. [Online]. Available: https://www.carritech.com/news/history-of-fiber-optics-in-telecommunications/
2. Kateeb, I., Alotaibi, K., Burton, L., Peluso, M. S., & Sowells, E. R. (2013, January). The fundamental component of telecommunications cabling. *Proceedings of the American Society for Engineering Education Pacific Southwest Conf*, 452–467.
3. Bellis,M.Thehistoryofthetelephone,ThoughtCo.Accessed:May5,2021.[Online].Available: https://www.thoughtco.com/history-of-the-telephone-alexander-graham-bell-1991380
4. Bruce, R. V. (2020). *Bell: Alexander Graham Bell and the Conquest of Solitude.* Lexington, MA: Plunkett Lake Press.
5. "The evolution of telephone cable," Copper Development Association. Accessed: Apr. 5, 2021. [Online]. Available: https://www.copper.org/applications/telecomm/consumer/evolution.html
6. Blom, J., Jonsson, B., & Kempet, L. (1994). Specification of Telephone Services. *Feature Interactions in Telecommunications Systems,* 197. Amsterdam: IOS Press.
7. Analog and Digital Transmission, Informit.com. Accessed: Apr. 6, 2021. [Online]. Available: https://www.informit.com/articles/article.aspx?p=24687&seqNum=5
8. Wikipedia contributors, Transatlantic telegraph cable, Wikipedia, The Free Encyclopedia. Accessed: Apr. 10, 2021. [Online]. Available: https://en.wikipedia.org/w/index.php?title=Transatlantic_telegraph_cable&oldid=1073596690
9. Marsh, A., The first transatlantic telegraph cable was a bold, beautiful failure, Oct. 31, 2019. IEEE Spectrum. Accessed: Apr. 8, 2021. [Online]. Available: https://spectrum.ieee.org/tech-history/heroic-failures/the-first-transatlantic-telegraph-cable-was-a-bold-beautiful-failure
10. Borth, D. E., Telephone – Overseas transmission, Encyclopedia Britannica. Accessed: Apr. 11, 2021. [online]. Available: https://www.britannica.com/technology/telephone/Overseas-transmission#ref1117813
11. Henrich-Franke, C. (2019). *Computer Networks on Copper Cables*, ComPuter Network Histories. Hidden Streams from the Internet Past. Zurich: Chronos.
12. Considerations for using hybrid copper-fiber cable, Jan. 12, 2021. Fluke Networks. Accessed: Apr. 15, 2021. [Online]. Available: https://www.flukenetworks.com/blog/cabling-chronicles/considerations-using-hybrid-copper-fiber-cable
13. 3 Benefits of using hybrid copper-fiber cable to provide data & power, Belden.com. Accessed: Apr. 20, 2021. [Online]. Available: https://www.belden.com/blogs/smart-building/benefits-of-hybrid-copper-fiber-cable-for-data-and-power
14. Miller, S. E. (1980). Overview of telecommunications via optical fibers. *Proceedings of the IEEE*, 68(10), 1173–1174.
15. Kao, C. K. (2010). Nobel lecture: Sand from centuries past: Send future voices fast. *Reviews of Modern Physics*, 82(3), 2299.
16. Wheen, A. (2011). *Dot-Dash to Dot. Com: How Modern Telecommunications Evolved from the Telegraph to the Internet*, Chichester: Springer.

6 Wireless Networks

ABBREVIATION

AAA	Authorization, authentication, and accounting
AMPS	Advanced Mobile Phone Service
ASAP	Aggregate Server Access Protocol
BSS	Basic service set
BSA	Basic service area
BSSID	Basic service set identifier
CDMA	Code-division multiple access
CSMA/CD	Carrier Sense Multiple Access/Collision Detection
DS	Distribution system
EDGE	Enhanced Data rates for GSM Evolution
ESS	Extended service set
ETACS	Extended Total Access Communication System
FCC	Federal Communications Commission
FTP	File Transfer Protocol
GPRS	General Packet Radio Service
GSM	Global System for Mobile Communications
HTTP	Hypertext Transfer Protocol
IBSS	Independent basic service set
ITU	International Telecommunication Union
LTE	Long-Term Evolution
MAC	Medium access control
MTS	Mobile Telephone Service
NMT	Nordic Mobile Telephone
NTACS	Narrow Total Access Communication System
P2P	Point-to-point
POP3	Post Office Protocol 3
PPTP	Point-to-Point Tunneling Protocol
SOFDA	Scalable orthogonal frequency-division multiple access
SSID	Service set identifier
SSL	Secure Sockets Layer
SNMP	Simple Network Management Protocol
SMB	Server Message Block
SMTP	Simple Mail Transfer Protocol
TDMA	Time-division multiple access
TLS	Transport Layer Security
UHF	Ultra high frequency
UMTS	Universal Mobile Telecommunication Service
WWAN	Wireless wide area network

DOI: 10.1201/9781003302902-6

WAP	Wireless access points
WiMAX	Worldwide Interoperability for Microwave Access
WLAN	Wireless local area network
WPAN	Wireless personal area network
WSN	Wireless sensor network

6.1 INTRODUCTION

Wireless networks have made life easier as the communication is using the electromagnetic transfer of signals between two or more end nodes without needing any physical medium or electrical conductor [1]. Radio communication is the most popular and common means of a wireless network. The lack of a wired tether is the main difference between a wired network and the wireless network, which is further in turn providing mobility to the end users or end stations of a wireless network. Moreover, the lack of tethered connection and mobility in a wireless network between the communicating devices present unique technical challenges. The mobility factor presented tremendous benefits as well but at the cost of certain drawbacks and technical challenges in terms of maintaining the wireless channel. The wireless signals are prone to the error due to channel noise, and the error probability also changes dynamically due to the dynamic change in the transmission medium and user mobility. For wireless network broadcast, where multiple stations can use the same medium, listening while speaking is somewhat complicated, and thus, coordinating the medium access control (MAC) is complicated. At the same time, network management becomes complicated as well due to equipment mobility and change in the network landscape. Battery energy has an important role to play in the wireless network to fulfill the energy expenditure of the user equipment. There has been the requirement of a radio resource management during radio transmission in certain range without interference to other transmissions. This has to further control the radio signaling and related protocols that will negotiate the optimal usage of network radio resources. Combining all the mentioned concerns of a wireless network, it was very hard to disentangle the responsibilities of different layers, like the MAC, routing, and resource allocation, in wireless networks.

The information transmits in the form of signals from the channel, and the communication between the sender and receiver can be viewed as happening through different layers of abstraction of the Open System Interconnection (OSI) model reference architecture. However, unlike the point-to-point (P2P) communication with wired networks, in general, the separation of functions between the layers is not cleaner as in the wireless networks. Figure 6.1 depicts a typical architecture of a wireless network; this diagram includes both a wireless cellular network and Wi-Fi.

The symbol stream with the transmission error rate, achieved by signals physical transmission and reception over the wireless channel, is the key service offering of the physical layer. The data link layer deals with the radio resource management and the network resource management, where the radio resource management has functions like power control, transmission rate allocation, and error control and the network resource management has functions like service scheduling and call admission control. The network layer deals with handoff management, location management, and traffic management and control. Radio communication has been the dominating

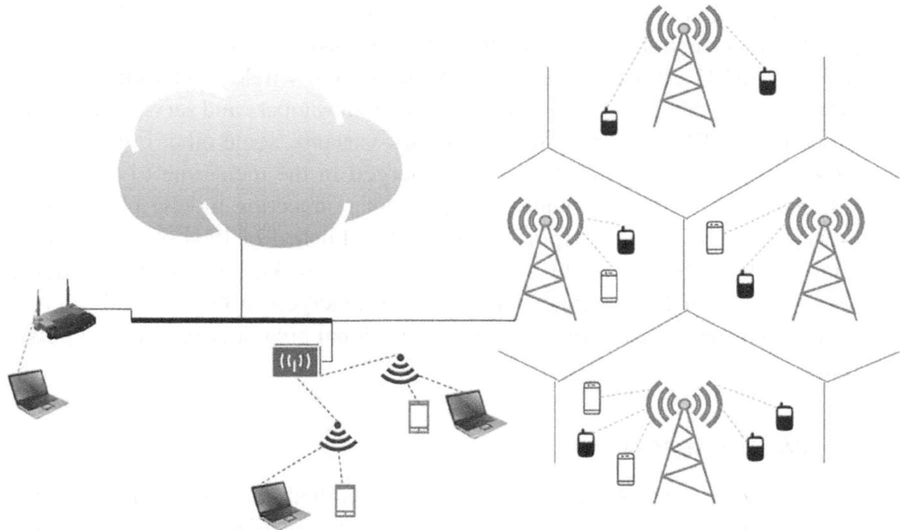

FIGURE 6.1 Diagram representing wireless communication network.

mode of communication for a wireless network. Different types of wireless networks are cellular, wireless local area networks (WLANs), wireless sensor networks, satellite communication networks, and terrestrial microwave networks.

6.2 WIRELESS NETWORKS AND COMMUNICATION

It has long been a scientific curiosity to transmit information wirelessly [2-5], without the need for a physical wired connection. Wireless telegraph using a separate telegraph line was developed to utilize signaling and audio communication using radio technology to communicate with ships and other moving vehicles. A complete wireless system was patented by Guglielmo Marconi in 1897; he gave a demonstration of wireless telegraphy for the first time [6]. It was based on long wave signals and spark transmitter technology, requiring very large and high-power transmitters. It also required long antennas to deal with the large wavelengths and that hindered its development for mobile system, except for maritime communications between the ship and shore.

We discussed in detail about the ALOHAnet in our earlier chapters; the origins of the radio frequency–based wireless network can be considered as the University of Hawaii (UH's) ALOHAnet research project in the 1970s. Since then, there has been a keen interest with the wireless technology and a growing demand for wireless connectivity. This demand was only able to fulfill by a narrow range of expensive hardware, proprietary technologies, less or no security mechanisms, and poor network performance from the 1970s through the early 1990s. Ratification of the Institute of Electrical and Electronics Engineers (IEEE) 802.11 standard in 1997 and subsequent development by Wi-Fi Alliance were the important events that led to the wireless networking becoming one of the fastest growing technologies of this century.

The IEEE 802.11 [7] standard was such a torch bearer to the development of wireless networks and starting point for a well-known brand which is famous as Wi-Fi. IEEE 802.11 primarily worked for the growth of wireless networking underlying technologies and gave focus to the equipment developers and service providers. Along with the IEEE standard wireless network variants, some other organizations have also contributed to technologies that emerged in the meantime to popularize wireless networks. Ericsson initiated research for connecting mobile phones with accessories, and the first IrDA specification was published in 1994. This led to the adoption of Bluetooth in 1999 by the IEEE 802.15.1 working group. Thereafter, a variety of wireless networking technologies have emerged so far to fulfill requirements based on the ranges of the data rate, operational area or coverage, and power consumption.

6.2.1 Wireless Telegraphy

Heinrich Hertz, a German physicist, had first identified and studied radio waves in 1886. He further produced, transmitted, and received Electro Magnetic waves (5 m–50 cm) using reflectors to concentrate the beam in 1888. Guglielmo Marconi, an Italian electrical engineer, was the first one who developed the practical radio transmitters and receivers around 1895–1896. Marconi soon developed a wireless system for transmission of wireless telegraphy signals [8] way beyond distances anyone could have predicted. It was his efforts toward commercialization of radio waves communication, which started around 1900. Marconi and Karl Ferdinand Braun were awarded the 1909 Nobel Prize for Physics for their contribution in wireless telegraphy.

Marconi was able to transmit and receive a coded wireless signal message at a distance of 1.75 miles near his home in Bologna, Italy, in 1895. He presented his developed operational wireless telegraph apparatus to the British telegraph authorities in February 1896 and filed his first British patent in June 1896. He then successfully sent wireless signals over a distance of 1.75 miles on Salisbury Plain in July 1896 with the assistance of Mr W.H. Preece, chief electrical engineer of the British Post-office Telegraphs. Furthermore, a greater distance of 4 miles was covered on Salisbury Plain with wireless signals in March 1897. Wireless communication covering a distance of 8 miles was established between Lavernock Point and Brean Down in England on May 13, 1897. His next demonstration was of a radio transmission to a tugboat covering an 18-mile path over the English Channel in 1897. In this year, Wireless Telegraph and Signal Company was founded as the first wireless company which bought most of Marconi's patents. The erection of the first Marconi station at Cape Cod, Massachusetts, started in 1900. Radio devices were installed by Marconi Company [9] at five stations on five islands of the Hawaiian group in March 1901. The Canadian government installed two stations in the Strait of Belle Isle this year. Also, the construction of the New York Herald stations at Nantucket, MA, and Nantucket light ship happened in 1901. The first international wireless message was sent by Marconi from Dover, England, to Wimereux, France, in 1899. Marconi further received the famous letter "S" at St Johns, Newfoundland, which was transmitted as a test signal from his English station on December 11, 1901. Several of the US government agencies, including the Navy, the Department of Agriculture, and the Army Signal Corps, started setting up their own radio transmitters in 1904. This was supported by a board with representatives from these agencies appointed by President Theodore Roosevelt to prepare recommendations for coordination of governmental development of radio services in 1904 [10].

6.2.2 Mobile Telephone Service – Precellular

The early voice communication as analog radio transmission was primarily driven by military applications. The experimental ship to shore radio telephone service was not initiated until 1919; however, the commercial radiotelephony begun in 1929 for passengers on ships in the Atlantic. A small and rugged mobile radio device was being installed in automobiles by this time, and the first operational mobile radio system was of the Detroit police in 1928; after that, the age of mobile radio had started. There were more than 5,000 radio-equipped police cars by 1934, which were operational for 194 municipal police systems and being served by 58 state police radio stations. Radio spectrum was always having shortage that could be used for practical radio systems. Also, at the same time for the available radio channels, mobile communication services were in competition with military and broadcast services. As a result, emergency and public service uses were assigned most of the mobile radio channels until the cellular revolution that started in about next 50 years.

An early Mobile Telephone Service (MTS) was launched in 1946 despite these difficulties and used 35 MHz channels; additional channels were also given further at 150 MHz, and later at 450 MHz [11]. Improved filtering and frequency stability techniques helped in narrowing these channels that eventually created more than 40 mobile telephony channels. The first commercial MTS was launched by Motorola in conjunction with the Bell System in the US in 1946. The early MTS was using a single high-power transmitter with analog frequency modulation techniques and was only able to provide service to limited customers. The early MTS was similar to the broadcast radio systems in that it used powerful transmitters from a high tower or rooftop and covered a distance of 20–30 miles. A separation of more than 50 miles was required between two transmitters to reuse a channel for different calls. The channels of one city were getting reused in another city, and about 40 simultaneous calls were limited to each city and its suburbs. The service demand was always increasing and resulting in severely overloaded channels with long waiting lists. In such scenarios, a practical method adopted was of giving preferential treatment to people with an important need for service, and the position in list of an average person may get worse over time. The earliest MTS had manual process of call placement with the help of a mobile operator placing calls in both directions. Automatic systems were introduced in the MTS with certain important advances in the 1960s, such as the addition of an "idle tone" to an idle channel. For many decades of existence, with manual and automated systems, the mobile telephony was conceived as a crowded but "elite" service generally unavailable to the public [12].

6.2.3 Mobile Communication – Cellular

The concept "cellular" was the breakthrough solution obtained to the issues in the earlier MTS of providing limited capacity service, where the idea was to replace a single high-powered transmitter with several low-power transmitters, and each will use a fraction of the total available spectrum to provide a coverage to a portion of the large service area. As long as the low-power base stations with the same frequency are sufficiently separated from one another, the channel frequencies could be reused across a large service area. Surprisingly, the origins of cellular was deep rooted in the past in 1947, when the first 150 MHz system was installed in St Louis. Then the Bell System (AT&T) had proposed 40 MHz for its implementation somewhere in the region between 100 and 450 MHz for a "broadband urban mobile system". The idea of allowing low-power radios in place of a high-power transmitter, serving a large geographic area using many small coverage areas, had already been put forward at Bell Laboratories at that time. Although conceived by Bell Labs in 1947, the Federal Communications Commission (FCC) denied this request, citing frequencies unavailability in that range. The required technology to implement the concept of cellular was not available until the 1970s. However, the essential components of the cellular system, like using small cells and the channels frequency reuse to increase the number of simultaneous calls per channel dramatically, were under discussion by this time. AT&T submitted a proposal for the cellular mobile concept to the FCC in 1971, and after more than a decade of deliberations, the FCC allocated 40 MHz of spectrum in the 800 MHz band in 1983. This was an important event which triggered the deployment of the first generation of commercial cellular systems [2].

6.2.3.1 First Generation

The first generation of cellular wireless systems development was led by the US, Japan, and parts of Europe. The first-generation cellular wireless networks were designed primarily for delivering of voice services using analog modulation schemes. The cellular concept with automatic switching and handover of ongoing calls was that technology in the evolution of wireless communication, which was something new and different from the earlier mobile communications systems. The world's first commercial cellular system, the first generation, was launched in Japan by the Nippon Telegraph and Telephone company initially in the metropolitan area of Tokyo in 1979. Nippon Telegraph and Telephone further expanded its network to cover the entire Japan's population and became the first nationwide first-generation network in a span of 5 years. The first cellular wireless system supporting automatic handover and international roaming was deployed in Europe, Denmark, Finland, Sweden, Norway, Austria, and Spain, by Nordic Mobile Telephone (NMT-400) system in 1981. However, Advanced Mobile Phone Service (AMPS) in the US and its variant Total Access Communication Systems (ETACS and NTACS) in Europe and Japan were more successful first-generation systems. With major difference of the channel bandwidth, these systems were almost identical to a radio standpoint, such that the AMPS had 30 KHz channel, whereas ETACS and NTACS used 25 and 12.5 KHz of the channel bandwidth, respectively [4].

6.2.3.2 Second Generation

The wireless revolution started in the early 1990s that led to the transition from analog to digital networks, which was enabled by advancements in metal-oxide-semiconductor field-effect transistor (MOSFET) technology. The first commercial digital cellular network was the second generation of mobile communication launched in Finland in 1991 and used the Global System for Mobile Communications (GSM) technology. The second-generation sparked competition in the mobile communication sector as the new operators were challenging the incumbent first-generation analog network operators. Various services provided by the second-generation system were text messages or Short Message Service (SMS) and picture messages or Multimedia Messaging Service (MMS) but not videos, and it was providing greater security to both the sender and receiver. To enhance the security feature, messages were digitally encrypted in the transit, allowing only the intended receiver to receive and read it. The methods of digital mobile access technology such as time-division multiple access (TDMA) and code-division multiple access (CDMA) were used, where signals to be assigned time slots under TDMA, whereas each user was to allocate a specific code to communicate over multiplexed physical channel under CDMA. The GSM standard was the most admired of all mobile technologies and was being used in most of the countries globally, making smooth international roaming service between mobile phone operators. The GSM standard was able to provide multiplexing of up to eight calls per channel based on TDMA in the 900 and 1,800MHz channel bandwidth, whereas in the US, a new block of spectrum was also auctioned in the 1,900MHz band by the FCC.

GSM technology was under continuous improvement to offer better services and quality to the global customer base during past 20years. That had led to the development of new technologies standard based on the original GSM standard, known as the 2.5 generation of mobile communication. The introduction the General Packet Radio Service technology was a step short of next mobile gencration and so is referred as 2.5 generation, to describe second-generation implementation with a packet-switched domain additional to the circuit-switched domain. This extension of the existing second-generation mobile network was to provide packet-based services to enhance the overall network speed and performance. The General Packet Radio Service had enabled services like web browsing, e-mail services, and fast upload/download speeds. Then came the 2.75 generation, the next subsequent enhancement, that is also known as Enhanced Data rates for GSM Evolution, evolving to the Enhanced Data rates for GSM Evolution networks with 8PSK encoding. The Enhanced Data rates for GSM Evolution was able to provide faster services than the 2.5 generation and gave faster Internet speed up to 128 kbps.

6.2.3.3 Third Generation

Third generation of mobile communication standards, similar to GSM, was introduced in the year 2000. The use of packet switching rather than circuit switching for data transmission is the main difference between third and second-generation mobile communications. Several features offered by the third-generation mobile are faster communication, high-speed web, increased security, video conferencing, 3D gaming, TV streaming, and so on. It is based on the International Telecommunication

Union formulated plan of 2,000 MHz band implementation globally to support a single and ubiquitous wireless standard throughout the world. The implementation plan of the International Telecommunication Union was called as International Mobile Telephone 2000 (IMT-2000) standard. The CDMA systems–based third-generation evolution led to CDMA 2000 standard, which is a family of third-generation mobile technology standards to send voice, data, and signaling between mobile phones and cell sites. Several variants of CDMA 2000 are based on IS-95 and IS95B technologies which were the second-generation standards also known as cdmaOne. The GSM-based third-generation evolution led to wideband CDMA, also called as the Universal Mobile Telecommunication Service.

6.2.3.4 Fourth Generation

4Fourth generation of mobile communication provides capabilities defined by the International Telecommunication Union in International Mobile Telecommunications (IMT) Advanced, providing significant improvement in the downloading speed of data up to 100 Mbps. Features of fourth generation include mobile web access, Internet Protocol (IP) telephony, gaming services, high-definition mobile TV, video conferencing, and 3D television. It is a common concept that every next decade, a new generation of mobile communication is being launched since the launch of first-generation mobile. The Long-Term Evolution standard first commercially deployed in Oslo, Norway, and Stockholm, Sweden, in 2009 and is getting deployed throughout most parts of the world since then.

6.2.4 Wi-Fi

Wi-Fi is the WLAN implementation of wireless Ethernet 802.11b standard referring to the technology supporting Internet connection wirelessly to wireless-enabled devices such as computers, laptops, and phones and has become the dominant wireless local area network (LAN) standard. Wi-Fi is very popular nowadays and has become a universal standard, and one can find Wi-Fi networks in offices, supermarkets, malls, hospitals, hotels, café, railway stations, airports, and many places, almost everywhere. Anyone can set up a Wi-Fi network with typically 100–500 ft of coverage with a high-speed Internet access because the Wi-Fi operates in unlicensed frequency bands. It has a major role in the digital disruption of businesses, with a variety of new service offering in the 15 years since the creation and commercialization of this wireless networking standard.

There were several wireless technologies used for connecting devices before 1999. Unfortunately, these different technologies were incompatible and were having certain limitations for wide implementations, so it was observed as the need of the time to have one such a cross-industry-recognized wireless LAN standard. For that purpose, the development started as an industry-recognized technical standard under specifications of IEEE 802.11, the new industry alliance that is now popular as the Wi-Fi Alliance [13]. This was formed as a new industry alliance and industry-wide organization of six major vendors for improving cross compatibility in 1999, the same way the development of Ethernet wired networking did in 1985. The new alliance was looking for a good name to promote the standards it was developing

TABLE 6.1
Table Mentioning Versions of Wi-Fi

Generations	Technology	Year
Wi-Fi 1	802.11b	1999
Wi-Fi 2	802.11a	1999
Wi-Fi 3	802.11g	2003
Wi-Fi 4	802.11n	2009
Wi-Fi 5	802.11ac	2014
Wi-Fi 6	802.11ax	2021

and hired branding consultants to find a catchier name than "IEEE 802.11b Direct Sequence". The name proposed by the consultants was "Wi-Fi" as a riff on "hi-fi", a word for audio and common to the world of music. The word Wi-Fi was catchy that caught on quickly, and it stuck. And the new industry alliance became the popular "Wi-Fi Alliance" that today a vast industries and technology vendors are part of it. The goal of this alliance is to promote wireless technology, protect Wi-Fi as a brand, and certify wireless products. Following the ratification of IEEE 802.11 and the Wi-Fi Alliance, almost every major networking company and computer hardware manufacturer started developing the Wi-Fi products. Apple had introduced the iBook G3 that was the first laptop integrated with Wi-Fi available for consumers in 1999. The Wi-Fi Alliance had more than 800 companies from around the world as of 2017 and had shipped over 3.05 billion Wi-Fi–enabled devices globally each year.

Wi-Fi generations are part of the marketing program with new customer friendly name introduction to the generations of Wi-Fi and based on technology physical layer (PHY) releases and are identified by a numerical sequence matching major PHY advancements. Several generations of the Wi-Fi are mentioned in the table below (Table 6.1).

6.2.5 WiMAX

Worldwide Interoperability for Microwave Access (WiMAX) [14] was created by the WiMAX Forum as a wireless broadband communication standards based on the IEEE 802.16 set of standards and is a wireless metropolitan area network (WMAN) technology. WiMAX systems provide multiple PHY and MAC options to deliver broadband access services in an economical way to residential and enterprise customers. WiMAX Forum was formed in June 2001 for the purpose of promoting conformity and interoperability and defining profiles of a predefined system to commercial vendors of the WiMAX standard, "WiMAX Forum" is to 802.16 as the "Wi-Fi Alliance" is to 802.11. Certification from the WiMAX Forum allows the vendors to sell WiMAX-certified fixed or mobile products and ensures a level of interoperability as long as they fit the same profile with other certified products. WiMAX has been described by the forum as "a standards-based technology enabling the delivery of last mile wireless broadband access as an alternative to cable and Digital Subscriber Line (DSL)" and can be considered as a wireless version of Ethernet to provide broadband access to customer premises intended primarily as an alternative to wired technologies (such as cable modems, Digital Subscriber Line (DSL), and T1/E1 links). Comparing to Wi-Fi, the WiMAX would operate similarly, but at higher speeds, covering large distances, and supporting more users. WiMAX can operate and provide good connectivity to those areas having limitations of traditional wired infrastructure and are difficult for reachability through a wired infrastructure.

WiMAX can provide several benefits in addition to providing 24 by 7 Internet access anywhere, such as can provide high data rates, high sector throughput, mechanisms of multiple handoff, mechanisms of power saving for mobile devices, advanced Quality of Service (QoS) and improved support with low latency to real-time applications, and advanced feature of authorization, authentication, and accounting functionality. WiMAX can be found today embedded in several devices, including USB dongles, Wi-Fi devices, laptops, and cellular phones.

WiMAX was designed initially to provide data rates of 30–40 Mbps and with the 2011 update can provide up to 1 Gbps data rates for fixed stations. The original version is IEEE 802.16 that has been specified for a PHY operating in the 10–66 GHz range; the version was updated to 802.16–2004 in 2004, with adding specifications for the 2–11 GHz band. Further development was in 2005 as 802.16e-2005 to use scalable orthogonal frequency-division multiple access as opposed to the fixed version using orthogonal frequency-division multiplexing with 256 subcarriers in 802.16d; thus, both of these Versions 802.16d and 802.16e are not compatible and need equipment replacement if an operator is to move to the later standard. The WiMAX Forum had approved officially an updated evolution roadmap to support the continued evolution of the WiMAX ecosystem in September 2013. This was based on recognizing the needs of service providers for flexibility to manage an ever-increasing demand for broadband data and to accommodate the harmonization and coexistence across multiple broadband wireless access technologies within a WiMAX Advanced network.

6.2.6 ZigBee

Zigbee is a wireless technology based on an IEEE 802.15.4 specification used to create personal area networks (PANs) with small, low-power digital radios, such as Internet of Things (IoT) [15] networks, home automation, medical device data collection, and other low-power low-bandwidth small-scale wireless connection needs. The name of the standard as "Zigbee" refers to the waggle dance of honeybees after their return to the beehive, and the concept was conceived in 1998, standardized in 2003, and was revised in 2006. Zigbee was developed as an open global standard to operate on the physical radio specification in unlicensed bands including 2.4 GHz, 900 MHz and 868 MHz. The 802.15.4 specification upon which the Zigbee stack operates gained IEEE ratification in 2003. This is a packet-based radio protocol specification intending low-cost, battery-operated devices, and the protocol allows devices to communicate in a variety of network topologies. The Zigbee Alliance is the organization responsible for the development of the Zigbee protocol; this alliance has members from over 300 leading semiconductor manufacturers, technology firms, and original equipment manufacturer (OEMs). Development of the Zigbee protocol was based on providing an easy-to-use wireless data solution characterized by secure, reliable wireless network architectures. The connection establishment and communication between Zigbee-certified products can take place using the same IoT language with each other, and millions of Zigbee products have already been deployed in smart homes and buildings. The concept of backward and forward compatibility has been kept while creating the Zigbee protocol. It has a defined data rate of 250 Kbps and is best suited for transmissions of intermittent data from a sensor or input device, secured by 128-bit symmetric encryption keys. Zigbee Pro created in 2007, also known as Zigbee 2007, where a Zigbee Pro device may also join and operate on a legacy Zigbee network and vice versa, operates using the 2.4 GHz Industrial, scientific and medical (ISM) band. Zigbee Pro 2015 can support self-forming and self-healing mesh network topology of network nodes up to 65,000 and is based on IEEE 802.15.4-2011 radio technology operating in 2.4 GHz (ISM band) of 16 channels (2 MHz wide).

6.2.7 Bluetooth

Bluetooth [16] refers to the standard for short-range wireless communication technology that is used to exchange data between fixed and mobile devices over short distances and build a PAN among the Bluetooth-enabled devices. Bluetooth uses ultra-high-frequency radio waves in the ISM bands, from 2.402 to 2.480 GHz, and was originally conceived as a wireless alternative to RS-232 data cables. The development of the "short-link" radio technology was initiated by Nils Rydbeck, Chief Technology Officer (CTO) at Ericsson Mobile in Lund, Sweden, in 1989. Intel, Ericsson, and Nokia, the three industry leaders, met to plan the standardization of this short-range radio technology to support connectivity and collaboration between different products and industries in 1996. Jim Kardach from Intel suggested Bluetooth as a temporary code name during this meeting; he later quoted, "King Harald Bluetooth…was famous for uniting Scandinavia just as we intended to unite the PC and cellular industries with a short-range wireless link". The name given to the technology dates back more than a millennium to King Harald "Bluetooth" Gormsson. The king was well known for uniting Denmark and Norway in 958 and has a dark blue/gray color dead tooth that earned him the nickname Bluetooth. The Bluetooth is managed by the Bluetooth Special Interest Group (SIG); SIG was launched with a total of five members: Ericsson, Intel, Nokia, Toshiba, and IBM in May 1998 and has over 35,000 member companies in the areas of communication, computing, networking, and consumer electronics. The Bluetooth has been standardized by the IEEE as the 802.15.1 standard, but IEEE no longer maintains the standard. The SIG looks the development of the Bluetooth specification, manages its qualification program, and protects the trademarks; to market a Bluetooth device, a manufacturer must meet Bluetooth SIG standards. The first consumer Bluetooth device was a hands-free mobile headset launched in 1999 and earned the "Best of show Technology Award" at Computer Dealers' Exhibition (COMDEX), whereas the first Bluetooth mobile phone was of the Ericsson T36 in 2001. IBM introduced the IBM ThinkPad A30 as the first notebook with integrated Bluetooth in October 2001. Bluetooth integrated circuit chips have been shipped approximately 920 million units annually as of 2009.

6.3 ARCHITECTURE OF WIRELESS NETWORKS

The architecture of wireless networks is similar to that of wired networks with the capability of the PHY to convert information signals into a form suitable for transmission through the air medium [17–19]. We will discuss concepts common to all types of wireless networks, with emphasis on its architecture. Different transmission media of wireless communication networks are infrared (Ir), broadcast radio, cellular radio, microwaves, and satellite link. Radio frequency– and Ir-based methods are the commonly used communication media of wireless communication. While designing the infrastructure of a wireless network, special consideration is given on several components that facilitate communications through an air medium using radio or light waves propagating.

As mentioned earlier that one of the most prominent benefits of deploying a wireless network is mobility, users tend to be mobile, constantly moving throughout a facility, campus, or city. The end devices can be laptops, portable computers, mobile phones, wireless printers, wireless headphones, and so on. The infrastructure of a wireless network has to interconnect network users with end systems, and it might consist of components such as base stations, access controllers, an application connectivity software, and a distribution system.

6.3.1 The OSI Reference Model

A reference model is the logical architecture of a network comprising the structure of standards and the protocols, with the goal of establishing connections between the network nodes and control the data flow in the network. The International Organization for Standardization (ISO) had developed the OSI reference model to provide a framework for the standardization of interconnecting devices. The model provides the conceptual framework that describes the logical operation of all networks, that include from a wireless PAN (WPAN) connection between a mobile phone and accessories to the global operation of the Internet. It is worth to take a note that the key features distinguishing the wired and wireless are defined at the data link layer (logical link control and MAC) and PHY of the OSI model which bring wireless networks to life. Figure 6.2 represents a reference OSI model architecture for the wireless network. The OSI model breaks the overall end-to-end connection (e.g., shown in the figure below, between Host A's application and Host B's application) into seven logical layers, where each layer has associated some task in the device communication.

The topmost layer of the OSI model or Layer 7 is the application layer that has standards to define provision of services to the applications. The applications layer services such as checking resource availability, and authenticating users. The protocol standards are Hypertext Transfer Protocol, File Transfer Protocol, Simple Network Management Protocol, Simple Mail Transfer Protocol, Post Office Protocol 3, and so on.

Below the application layer is Layer 6, or the presentation layer of the OSI model that has standards for the presentation of data to and from the application layer and controls the translation of data from one presentation format to another. The protocol standards are Secure Sockets Layer, Transport Layer Security, and so on.

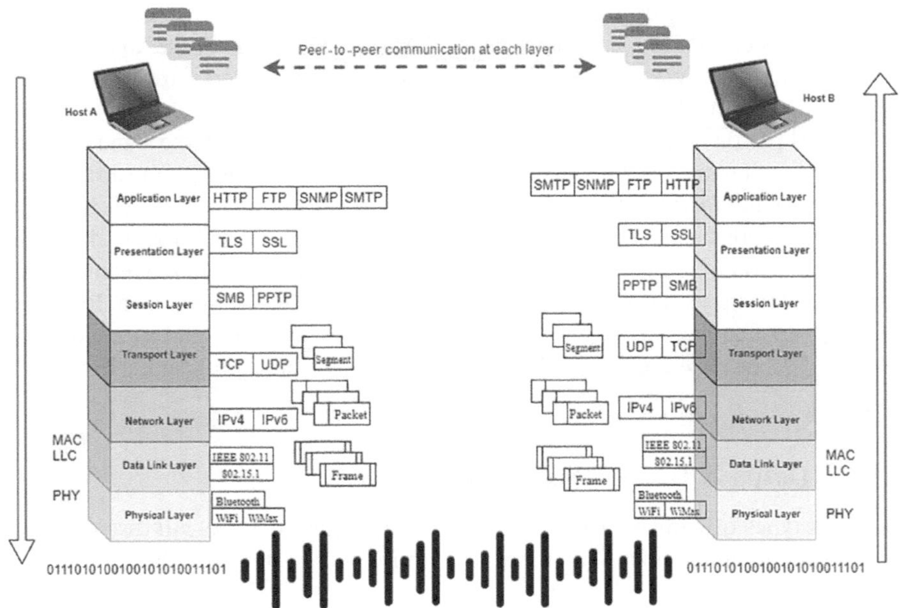

FIGURE 6.2 OSI model reference representing for a wireless network.

The layer below the presentation layer in the OSI model is Layer 5 or the session layer. This layer has standards that manage communication between presentation layers of the sending and receiving devices. The session layer establishes, manages, and terminates the sessions. Protocol standards of the session layer are Point-to-Point Tunneling Protocol, Server Message Block, Aggregate Server Access Protocol, and so on.

Layer 4 of the OSI model is the transport layer that has standards for reliable end-to-end data transfer between the devices of the network. It has a mechanism for error recovery, data flow control, segmentation, and reassembly of data and can provide a connection-oriented and connectionless transport of data. Protocol standards of this layer are Transmission Control Protocol (TCP) and User Datagram Protocol (UDP).

Below the transport layer is Layer 3, the network layer of the OSI model, that has standards to define the management of a network connection, routing, packet relaying, and connection termination between nodes. The protocols are IPv4, IPv6, and so on.

Layer 2 of the OSI model is the data link layer that specifies the way to access devices, share the transmission media, and ensure the reliability of physical connection. The two important mechanisms under the data link layer are MAC and logical link control. The wired network has an important feature of collision detection part of Carrier Sense Multiple Access/Collision Detection (CSMA/CD) under the MAC sublayer that is not possible for radio access wireless networks. For the wireless networks, a variant of CSMA/CD known as CSMA/collision avoidance is used where collision detection is not possible. CSMA/collision avoidance has some similarities with CSMA/CD apart from no collision detection by the transmitting device, which is the devices can sense the medium before transmission of the packet and can wait if the medium is busy. Protocol standards for the wireless networks are Wi-Fi (IEEE 802.11), Bluetooth (IEEE 802.15.1), and so on.

The PHY is the lower most layer of the OSI model dealing with the standards that control the transmission of data stream over a particular medium. Transmission of the data is in the form of electrical signals/pulses, beams of light, or the electromagnetic underneath this layer. It is that layer which is responsible to provide the lower layer foundation of wireless networks. Protocol standards covering the technologies of the wireless networks are Wi-Fi, Bluetooth, WiMax, Zigbee, and so on. Based on the type of technology used (e.g., Ir, radio frequency, near field) and its application (PAN, LAN, or wide area network [WAN]), the range and significance of the PHY vary.

6.3.2 Different Types of Wireless Network

Different types of wireless networks are shown in Figure 6.3; we will discuss about them in detail with various equipment and required connections for them.

6.3.2.1 Wireless Wide Area Network

Wireless WAN is a networking technology or concept that provides regional, nationwide, and global wireless network coverage and thus offers a wider network connectivity like a WAN. Technologies of mobile communications are used to encapsulate the network traffic in a wireless WAN for providing access to the user outside the range of a wireless local or wireless metropolitan network. The examples are Worldwide Interoperability for Microwave Access (WiMAX) [14], Universal Mobile Telecom System, CDMA 2000, GSM, or third-generation networks, and so on. Also, such networks enable their users to make a phone call through a wireless WAN or a wired telephone system, and users can also get access to the Internet for web-based applications.

Generally, mobile towers are located nearly everywhere or in most of the parts of many countries that provide cellular network coverage. Whenever a user wants communication, a connection request is then routed to the nearest mobile tower where the customer is located. Based on that, the user equipment gets connected

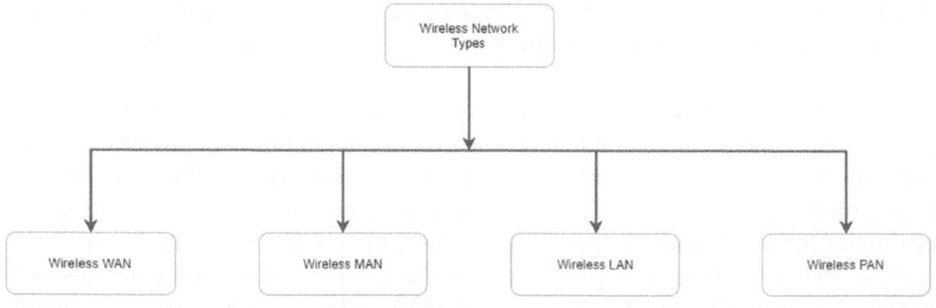

FIGURE 6.3 Different types of wireless network.

through the base transceiver station (BTS) that in turn is connected either to the wired Internet directly or to another tower connected with wired Internet connection.

Some of the characteristics of a wireless WAN are that the rates of transmission greatly reduced compared with the physical connections. WiMAX technology is based on IEEE 802.16 standards for broadband wireless access with an acceptable range of 70 Mbps transmission rates at approximately 30 miles. Throughput of such network decreases as distance increases and vice versa. This networking technology is ideal for those users who are away from home and need connectivity virtually anywhere in the area of coverage with mobility in a car, an airport, or a sporting event, and so on.

6.3.2.2 Wireless Metropolitan Area Network

A WMAN is a fixed wireless network that interconnects buildings or locations and has been installed in cities around the world to provide access for those people who are outside an office or home network. The WMANs are generally of two basic types: the back haul and the last mile. The back haul network is generally preferred by enterprises when they do not want to install or lease fiber to connect their facilities over a large campus or city. A back-haul–based WMAN is achieved by cellular-tower connection and Wi-Fi hotspots. On the other hand, the last-mile–based wireless WAN solutions could establish a wireless connection as an alternative to residential broadband DSL/cable modem.

The WMAN has a wider coverage area than office or home networks with the same principles. Wireless access points (WAPs) are installed on the sides of building, infrastructure, or poles throughout the coverage area to give coverage access to the users. These access points (APs) broadcast a wireless signal throughout their area of coverage and are then connected to the Internet via a wired connection. Users of a WMAN are connected to the nearest AP for connection to their desired destination, and the AP then forwards the network traffic (packets) through its wired connection to Internet.

6.3.2.3 Wireless Local Area Network

A WLAN is a wireless network used to link two or more network devices wirelessly within a limited area such as a home, school, laboratory, campus, or office to form a LAN. This provides users mobility as an added advantage over the wired LAN so that the user can move around within the coverage area and be connected to the network. Using an Internet gateway, a WLAN user can be provided Internet connection. WLAN technology is very popular now as earlier it was used within offices and homes and now is preferred almost everywhere within premises or buildings for wireless network access such as in the shops, stores, shopping malls, and restaurants. At the same time, COVID-19 pandemic has forced office workers, students, teachers, and others to work and study from home and that has greatly increased the use of wireless home networks.

Most of the modern WLANs are IEEE 802.11–based standards that are marketed under the Wi-Fi brand name; the home networks are generally fulfilled by a wireless router connected to modem broadcasting a wireless protocol based on 802.11

standards, whereas the office networks are more complicated as compared to the home networks using the APs mounted on the ceiling to broadcast wireless signals within the surrounding area. Accomplishment in the large offices may require multiple APs, connecting each of them over a backbone network through a wired connection to distribution switch.

6.3.2.4 Wireless Personal Area Network

A WPAN is that wireless network which covers a very limited area, used for interconnecting electronic devices centered on an individual's workspace proximity, limited to a room, office cabin, and so on. A WPAN is used for communication among the devices themselves and can be used for connecting a higher level network or Internet by using one master device working as a gateway. Typically, it can be ranging between 10 m and maximum 100 m of coverage distance for most of the applications. A WPAN is a low-powered, short-distance wireless network technology based on wireless protocol standards such as Bluetooth, Zigbee, Ir, and Wireless USB. Bluetooth communicates over short-range radio waves and can be used to connect Bluetooth devices such as keyboards, pointing devices, audio headsets for hands-free calls/audio experience, printers, smartwatch, phones, or computers. A Zigbee technology is used for connecting stations in an IoT network. Ir technology is very limited and used for the line-of-sight communication between the devices, such as connecting AC, TV, or several other appliances through a control on remote device.

6.3.3 Wireless Network Topologies

Wired network implementation has an important consideration of network topology, specifically, a LAN, where the network topologies represent the way of physical connection of all the devices in the LAN. Selection of the most suitable network layout is vital for the efficient network operation. Like the wired networks, the wireless networks implementation as well has topologies. These topologies are important consideration for wireless networks because of factors such as latency, power, speed, and redundancy. Modern networks consist of various types of devices that may be router, switch, smartphone, Bluetooth-enabled devices like headphones, smart bulbs, smartwatches, and many other devices. Each of the devices is a node in the network, and the topology describes how it connects and communicates in a network.

The important wireless network topologies are listed below.

6.3.3.1 Point-to-Point Wireless Network

The P2P wireless topology is the simplest wireless network architecture using a wireless radio link to connect two nodes and can communicate directly with each other. The range of the P2P link can be from a short-range link with two locations few hundred meters apart up to a long-range wireless link connecting two locations tens of miles distant to each other. Distance of a P2P wireless link is affected by factors such as height of both communicating radio devices, carrier frequency, signal power, and interference (Figure 6.4).

FIGURE 6.4 Point-to-point wireless network.

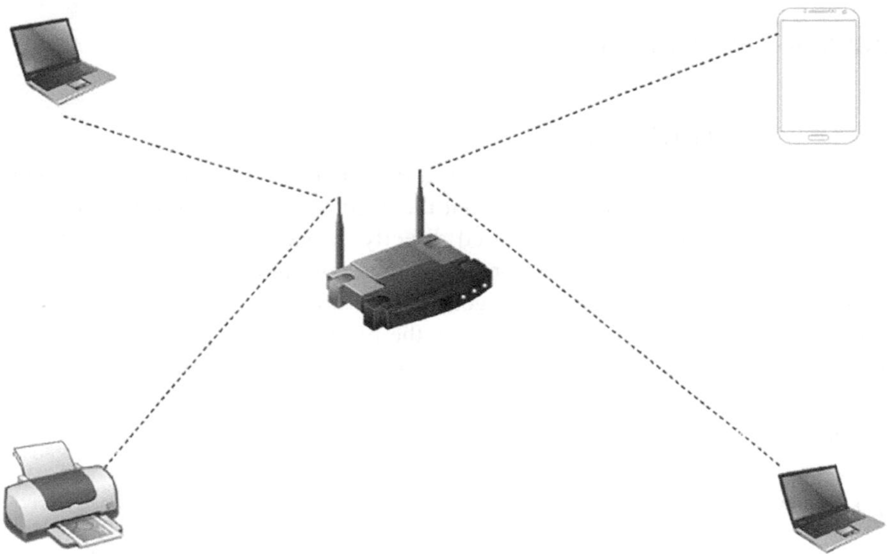

FIGURE 6.5 Wireless star network.

6.3.3.2 Star Wireless Network

A star wireless topology is one of the most commonly used wireless networks. A star network has a central gateway node, and all the other end nodes are connected to that gateway or central node. The node at the center of a star network plays an important but role similar to the hub in a wired network. Each of the end nodes while communicating to each other has to reach only through the gateway node. In a star network, the repeaters are also used sometimes to fill gaps in the coverage (Figure 6.5).

6.3.3.3 Tree Wireless Network

A tree network can be considered as the extension of the star topology that is used in bigger networks compared with the star network. In Figure 6.6, we can see that the three different star networks from three different central gateway devices are connected to AP device like branches of a tree (Figure 6.6).

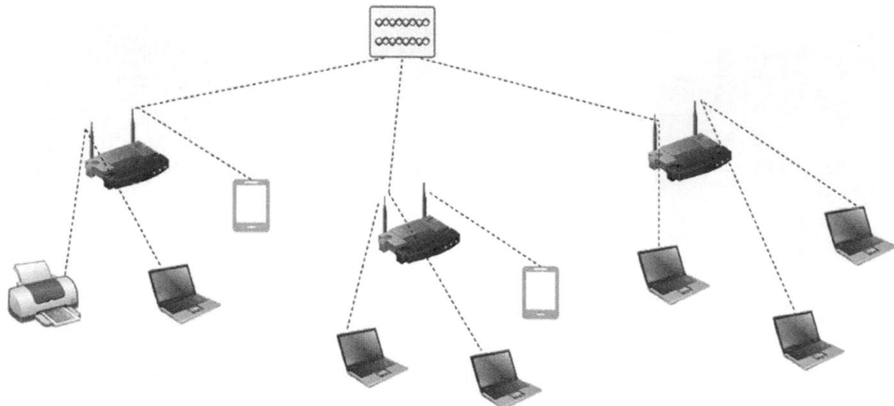

FIGURE 6.6 Wireless tree network.

6.3.3.4 Mesh Wireless Network

A mesh wireless network is such a network in which mobile nodes can communicate directly with adjacent nodes without the need for central controlling devices. Therefore, all the nodes are connected directly to each of the other nodes in a full mesh network topology. Functionality of data routing is distributed throughout the entire mesh network rather than getting controlled by one or more dedicated devices. Some of the networks also follow the partial mesh topology, where some nodes are connected to some of the other nodes, but other nodes are connected only to those nodes or gateways with which they have to exchange most of the data (Figure 6.7).

6.3.4 Wireless Service Modes

A wireless network service mode refers to a state or means of operation that a network node uses while connecting and communicating with other nodes in the wireless network. A service mode describes various operational characteristics of the wireless signals used and also impacts the overall performance of network.

Some terminologies used in the wireless service modes are as follows:

Basic Service Set (BSS): A BSS consists of one WAP interconnecting all associated wireless clients. A WAP is a piece of hardware called an AP or WAP that works as a gateway between the wireless and wired backbone. All the wireless clients (or BSS clients) have to start communicating with the WAP first within its range before communicating the wired network. After WAP connection is established, the wireless client can start communication between other clients or the attached wired network with flow of packets at Layer 3 (routed network) or Layer 2 (switched network).

Basic Service Area: A basic service area (also called a cell) is the area of coverage bounded by the reach of the wireless signals of a WAP.

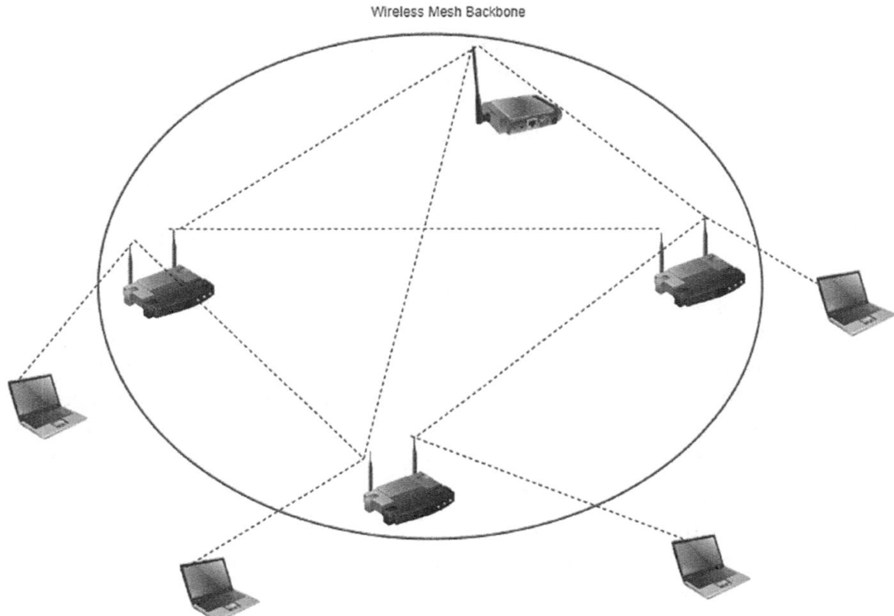

FIGURE 6.7 Wireless mesh network.

Basic Service Set Identifier (BSSID): A BSSID is a machine-readable identifier that is unique for the WAP. A BSSID is usually derived from the WAP's wireless MAC address, and hence, it is in the format of a MAC address.

Service Set Identifier (SSID): A SSID is a human-readable identifier that is nonunique for WAP. It is used by the WAP for advertising its wireless service.

Distribution System (DS): Usually like a distribution layer in a wired 3-tier architecture of the network design (comprising core layer, distribution layer, and access layer switches), the WAPs are also connected to the wired network infrastructure such as Ethernet in the backbone network that is called as a DS. The translation of frames between 802.3 Ethernet and 802.11 wireless protocols is performed by the WAP with a wired connection to the DS.

Extended Service Set (ESS): The ESS is the extension of the coverage provided by a single BSS, when a single BSS provides insufficient coverage. Two or more BSSs are joined to form a union, interconnected by a wired common DS to form an ESS. Each ESS has its SSID for identification, and each BSS uses its BSSID for identification.

Below are the two service modes in wireless networks described by the IEEE 802.11 standard.

- **Infrastructure Mode:** The infrastructure mode of a wireless network facilitates the wireless clients to get connected with the WAP. Clients, such as laptops and smartphones, are connected with the WAP for the wireless communication, and most of the Wi-Fi networks are deployed in the infrastructure mode. The WAP also sometimes has a wired connection or a permanent wireless connection with other WAPs. WAPs are generally fixed location wise to provide service to their client nodes within their coverage area, whereas some of the wireless networks may have multiple WAPs connected to provide service using the same SSID and security arrangement. In such cases, client nodes can connect to any of the WAP to join the network based on client systems logic to select the most appropriate WAP giving the best service, such as WAP offering the strongest signal. The infrastructure mode configuration also allows wireless clients to roam between basic service areas with the WAPs sharing the same SSID.

- **Independent BSS:** The independent BSS is also referred to as an ad hoc or peer-to-peer wireless network service mode. This is called as an independent BSS due to the fact that, in this mode, no hardware WAP is required to form a BSS. And thus, it is an independent BSS because ad hoc networks do not rely on any device other than the stations themselves. A BSS is formed as soon as wireless devices or "stations" are connected to each other over a wireless network. It allows the communication between the wireless nodes or peers if any of the network node is within the range of another node (peer) and both nodes agreeing on some basic common parameters. At the time of forming connection, the first node defines the radio parameters and connection name, whereas the other node has to detect the connection and adjust its own parameters to connect to the first station. Also, if one of those network nodes is having a wired connection to some other network, it can provide access to that network as well.

  These networks are limited in functionality because of the fact that no central device is present to decide on common parameters (such as radio parameters, priority, coverage, and what happens if the first station disappears). Usually, the WAP is treated as a central device in the infrastructure mode to organize the communication that defines common sets of parameters, such that the WAP organizes the BSS and wireless clients send their signals to the WAP, which then relays the signal to the destination wireless node or the wired network.

6.4 CONCLUSION

The wireless communication has created revolution in fundamental changes to data networking and telecommunication. It has the provision of freeing the user from the cord through numerous wireless technologies available. Radio communication is the most popular and common means of a wireless network. The lack of a wired tether is the main difference between a wired network and the wireless network, which is further in turn providing mobility to the end users or end stations of a wireless network.

REFERENCES

1. Wikipedia contributors, Wireless network, Wikipedia. Accessed: May 5, 2021. [Online]. Available: https://en.wikipedia.org/w/index.php?title=Wireless_network&oldid=1071389820
2. Hindle, P. (2015), History of wireless communications. *Microwave Journal.* Accessed: May 8, 2021. [Online]. Available: https://www.microwavejournal.com/articles/24759-history-of-wireless-communications
3. Nassa, V. K. Wireless communications: Past, present and future. *Dronacharya Research Journal*, 3(2), 50-54.
4. Sarkar, T. K., Mailloux, R., Oliner, A. A., Salazar-Palma, M., & Sengupta, D. L. (2006). *History of Wireless.* Hoboken, NJ: John Wiley & Sons.
5. Seymour, T., & Shaheen, A. (2011). History of wireless communication. *Review of Business Information Systems (RBIS)*, 15(2), 37–42.
6. Marconi, G. (1909). Wireless telegraphic communication. *Nobel Lecture*, 11, 198–222.
7. Hiertz, G. R., Denteneer, D., Stibor, L., Zang, Y., Costa, X. P., & Walke, B. (2010). The IEEE 802.11 universe. *IEEE Communications Magazine*, 48(1), 62–70.
8. Marconi, G. (1899). Wireless telegraphy. *Journal of the Institution of Electrical Engineers*, 28(139), 273–290.
9. Wikipedia contributors, Marconi Company, Wikipedia. Accessed: May 6, 2021. [Online]. Available: https://en.wikipedia.org/w/index.php?title=Marconi_Company&oldid=1062397895
10. Greenstein, L. (1999). 100 Years of Radio. Speech at WINLAB Marconi Day Commemoration, Red Bank, NJ, September, 30.
11. Borth, D. E., Mobile Telephone, Oct. 11, 2017. Encyclopedia Britannica.
12. Dubendorf, V. A. (2003). *Wireless Data Technologies.* Hoboken, NJ: John Wiley & Sons.
13. Wi-fi alliance, Wi-fi.org. Accessed: May 15, 2021 [Online]. Available: https://www.wi-fi.org/
14. WiMAX Forum, Wimaxforum.org. Accessed: May 20, 2021 [Online]. Available: https://wimaxforum.org/
15. Building the foundation and future of the IoT, CSA. Accessed: May 20, 2021. [Online]. Available: https://zigbeealliance.org/
16. Bluetooth technology website, Bluetooth® Technology Website. Accessed: May 21, 2021. [Online]. Available: https://www.bluetooth.com/
17. Wong, L. Y. K., Louie Wong, Louiewong.com. Wireless Network Architectures. Accessed: May 10, 2021. [Online]. Available: https://www.louiewong.com/archives/407
18. Jacobs, D., The 4 different types of wireless networks, Jan. 5, 2021. SearchNetworking. Accessed: May 12, 2021. [Online]. Available: https://searchnetworking.techtarget.com/tip/The-4-different-types-of-wireless-networks
19. Johnson, A., Wireless Concepts, Ciscopress.com. Accessed: May 12, 2021. [Online]. Available: https://www.ciscopress.com/articles/article.asp?p=2999384&seqNum=4

7 Circuit Switching and Packet Switching

ABBREVIATIONS

ANI	Automatic number identification
ARPANET	Advanced Research Projects Agency Network
CCITT	Consultative Committee for International Telephony and Telegraphy
DTMF	Dual-tone multiple frequency
NPL	National Physical Laboratory
PABX	Private automatic branch exchange
PSTN	Public switched telephone network
DID	Direct inward dialing
VC	Virtual circuits

7.1 INTRODUCTION

We have come across the data transmission technologies over communication links in a network over wired as well as wireless mediums. Communication networks are generally having different types of communication links. For a direct connection of two nodes in a network, the communication links will be a point-to-point link, whereas for multiple nodes, the connection will be through a multiaccess or broadcast link. For those situations, when we find the nodes/stations are not having a direct link or connection to each other, we need some sort of mechanism to have the communication setup between those nodes. That mechanism is broadly referred to as the switching concept in the communication network. In this chapter, we are going to examine the technology of circuit switching and the packet switching, whose evolution is a remarkable achievement in the history of telecommunication that has taken place over the years. The switching concept forms a very important component of a communication network. It refers to the mechanism of connecting the incoming path to the desired outgoing path and directing the messages to reach the destination.

So, the concept of switching refers to the technologies used, where the message from one node to be relayed through intermediary nodes at different network junctions and will finally reach to the destination node, and data transfers from source to destination through a series of nodes. These intermediary nodes are electronic devices, known as the switches of the communication network, and have a significant role in the establishment and overall operation of the network. The electronic system of switches and links and the controls governing their operation to allow data transfer and exchange among multiple users form an integral part of network communication. A network is established by linking switches with nodes or stations, such that each user is connected to one of the nodes. Each of the links in the network refers a

DOI: 10.1201/9781003302902-7

communication channel that may be wired (copper cables, optical fiber cables, and so on) or radio waves for wireless channels.

7.2 SWITCHED NETWORK

Popular switching techniques in the computer networks are commonly used in two different ways, circuit-switched or a packet-switched networking. Circuit switching [1–3] is one of the switching technologies and concepts that has been the dominant technology for voice communications since the invention of telephone in late 1800s. It has also been used up to some extent for the data communication, during the era of computer evolution. The circuit-switched network has to establish a dedicated physical connection through the network that will be held connected until the communication is necessary; the traditional (analog) telephone system is an example of this network (Figure 7.1).

The telephone system with the possibility of long-distance voice communication was invented and demonstrated by Alexander Graham Bell in March 1876 [4]. That demonstration was of the point-to-point communication, where a calling party chooses

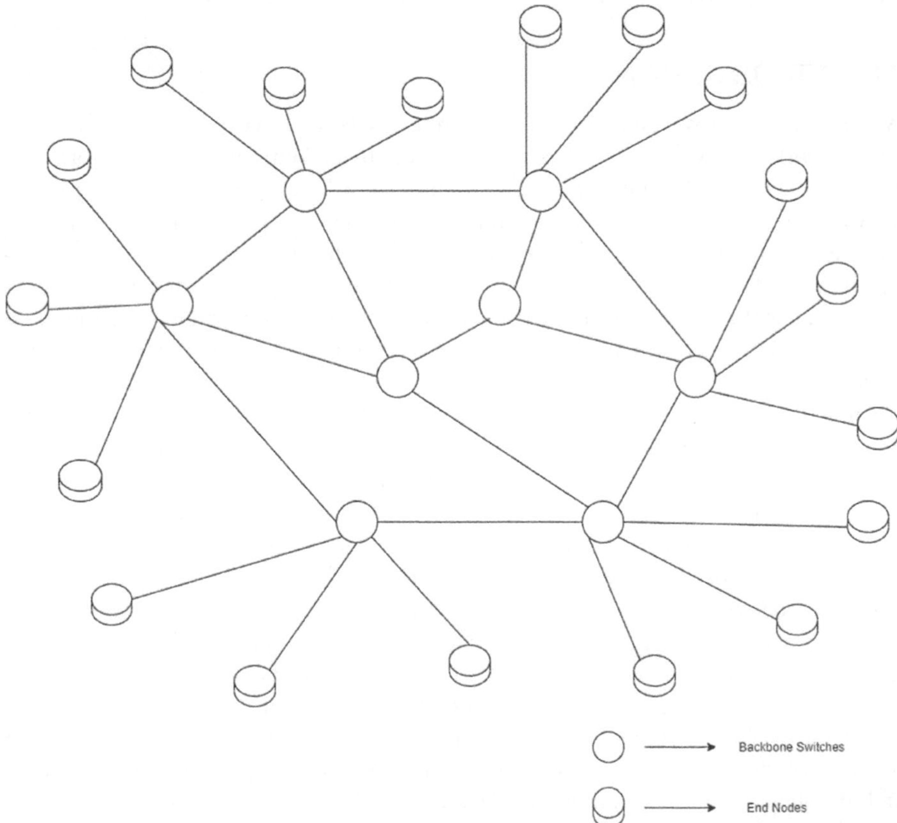

FIGURE 7.1 Switched telecommunication network.

the suitable link to establish a point-to-point connection with the called party. And the telephone system has further requirement of some kind of a signaling mechanism along with the voice communication between the two parties. To establish the point-to-point connection, both of the parties were supposed to be linked with wires to connect their telephone sets at both the ends. Considering that, if the number of telephone sets is less in a geographic location, then the complexity of establishing such a connection would be reduced. On the contrary, the connections will be more complex and to be fulfilled by a mess network if the number of telephones sets is large. A mess network with "n" number of nodes or telephone users will require $n(n-1)/2$ total number of point-to-point links on the network. The connection establishment as well as the operation and maintenance of a mess telecommunication network with higher nodes will be very complicated for the telecommunication service provider. This led to the birth of the switched telecommunication network, and the technique of switching used was what became known as the "circuit switching". A switching system in the telecommunication network was introduced between the connecting nodes, with the capability of switching to establish a connection as and when needed, where a switching office was recommended by Alexander Graham Bell to perform switching and to maintain the telephone connections.

On the other hand, a packet-switched networking has to route data in the form of packets to proceed independently within the network. Packet switching was another switching technology that was developed in 1970s and came into limelight with the development of the Advanced Research Projects Agency Network. Each packet has to be stored temporarily at each intermediate node during transit and then to be forwarded to the next node in a process called store and forward. The packet switching is now considered as the most effective means for data communications and is also said as the switching technologies of the Internet.

Figure 7.2 represents different types of switching in telecommunication networking. Another switching technique that was the precursor to packet switching was the message switching, where individual messages in their entirety are separately switched along their path. This switching technique has no requirement of establishing circuits exclusively for the messages from source to destination; messages are transferred based on store-and-forward mechanism instead. The implementation of message switching is often performed nowadays over packet-switched or

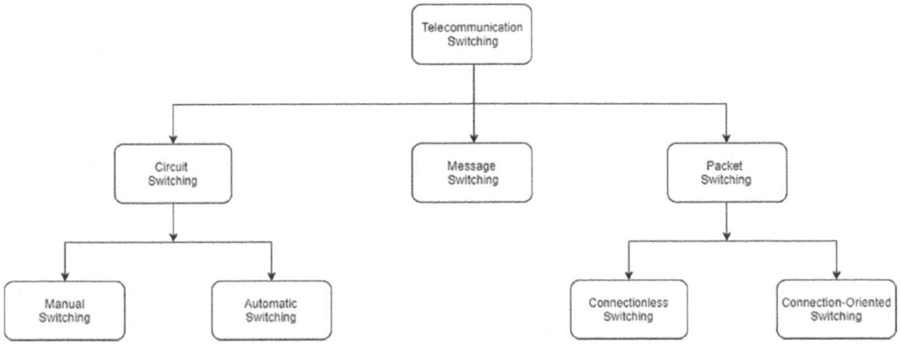

FIGURE 7.2 Types of switched telecommunication network.

circuit-switched data networks, where each message is treated as a separate entity and contains address information. That address information contained in the message is read at each switch, and the transfer path is determined based on that. Transfer of several messages might not always take place over the same path toward the destination. Each message has to be stored first, before transmission to the next switch, and thus, the transmission mechanism is called the "store-and-forward" method. In the switching operation of message switching, each of the switching devices has to wait until it receives the entire message. After receiving the complete message, that switch then stores and forwards over the link to the next switch, and this approach is continued until the message is received at the destination.

Packet switching is basically a specific message switching type, where the message is broken down into smaller pieces called as packets. As a result, packet switching allows the switching devices to start transmission as soon as the first packet of message arrives. The packet-switching technique came into existence after advancement of computer communication and data networking. One of the benefits it provides is of saving time, specifically when there are more numbers of hops between the source and destination. As with more numbers of hops, the path also has more propagation delay, and the minimum switching time taken by a device for the switching action will be more suitable.

7.3 CIRCUIT SWITCHING

Circuit switching [5] is a switched network implementation of telecommunication, where two nodes have to establish a dedicated communication channel or the circuit connection prior to the communication of nodes. The circuit switch has to connect one circuit's output with another's input to let the signals flow through that circuit connection. Nodes have full real-time access to each other up to the bandwidth of the circuit in a circuit-switched network; hence, it guarantees full channel bandwidth and remains connected throughout the session. These circuits were mostly point to point in the network, generally connected over a physical facility such as a pair of cables before 1891 and was needing a manual operator's effort to connect the circuits. Further it was until Almon Strowger, who invented the replacement of telephone operator manual process in the form of an electromechanical switch, and created the first automatic switching system. The circuit-switch operation looks like the nodes were physically connected with an electrical circuit.

The first telephone switch started working in New Haven, Connecticut, in January 1878. Since then, the switching technology had advanced drastically over the span of the time with keeping the basic functionality similar, where the basic feature of a circuit-switched network is to establish a physical circuit connection between users over telephone lines [6]. Every telephone network has switches to perform the circuit-switching activity of telephone lines, or establishing of the circuit connection, by physically connecting telephone lines. In such a way that the interconnection of the telephone lines can be understood in simple terms, both parties involved in the call, the called party and the other being the calling party, were supposed to be connected via a circuit connection on the same switch. The connection of the telephone lines happens in such a way that the caller first dials the number of the desired called party, followed by line availability check by switch, and if the lines are available, the two lines are to be interconnected by the switch. The connection of the lines is maintained until one party hangs up the call, and at that time, the switch has to terminate the circuit connection, and both lines are freed for other calls.

So, the characteristics important to this type of switching or the "circuit switching" are mentioned in the below points and were of significant consideration while developing a circuit-switched telecommunication network.

- The circuit between the two parties has to be created first before they can talk. To make the connection, the switch takes some time to check if a connection can be made for them.
- When a connection is established, a dedicated circuit connection is established until that connection is terminated.
- Three, one accounting policy telephone companies implemented due to the high cost of switches. This was done for the purpose of recovering their investment by instituting a minimum charge for each telephone call (Figure 7.3).

However, some issues were found with implementing the circuit-switching technique in computer communication or for the data network and was generally not observed as the suitable mode of communication. Some of the issues were generally understood earlier and were once again strongly confirmed with the experiments by Roberts and Marill in 1965. It also became evident with the experiments that the circuit-switching architecture of the telephone system was not the only problem for computer communications. There was also a search for a suitable communication system to existing host operating system software design as there was only one host as a master system, and all the other connecting devices were its slave system. Peer-level computing did not exist at the time, and hosts were not designed to recognize or interact with peer-level computers.

Fortunately, Paul Baran, an inquisitive innovative scientist at RAND Corporation [7] and a pioneer in the development of computer networks, had already explored the problems of circuit switching and started working in this area. He was one of the two

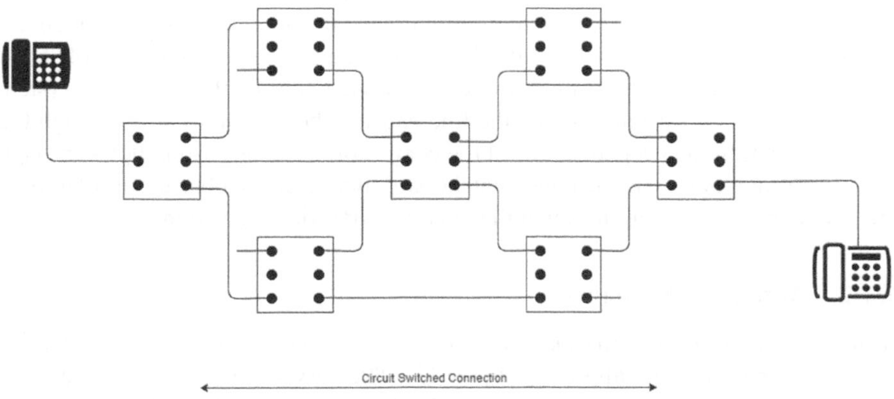

FIGURE 7.3 Telephone – circuit-switch network.

independent inventors of packet switching, made his concept of the message-based communication system publicly known by 1962. Independently, Donald Davies, an English scientist, also reached at the same conclusions as Baran in 1965 and coined the name which is popular today as "packet switching".

7.3.1 Manual Switching

As we have mentioned in previous sections, the switching systems in the early stages of telecommunication were manually operated, which requires the operator's manual intervention at the telephone exchanges to establish connections between two communicating parties. With later advancements further in telecommunication, automatic switching systems were introduced to minimize the disadvantages of manual switching operation [8–11]. Manual exchanges are not operational nowadays in the public switched telephone network. However, in some cases, operator assistance is still required at a private automatic branch exchange on a routine basis to connect the incoming calls with the required extension numbers. This particular requirement of manual switching is also getting decreased with large scale introduction of direct inward dialing facility in most of the private automatic branch exchange.

The operation of manual switching service was based on the customer lifting the receiver off-hook and asking the operator at the manual exchange to connect the call to the called party's number, where the number should be in the same central office and be located on the operator's switchboard. To connect the call, a manual exchange operator was supposed to plug the ringing cord into the jack corresponding to the called party line. For those conditions where the called party's line was present on a different switchboard in the same office or in a different central office, the operator was supposed to plug the ringing cord into the destination switchboard or office trunk and ask the "B" operator to connect the call with the called party.

Common battery service is mostly provided by the urban exchanges, where the central office has to provide the power for operation of the transmitter and automatic dials signaling to the subscriber telephone circuits. The wires used to connect subscriber's telephone with the exchange contain a nominal 48V DC potential across the conductors from the telephone service provider end in a common battery system. The circuit connection of a telephone system is such that it represents an open circuit when the telephone is on-hook or idle. The moment a subscriber's phone is off-hook or removed from its idle position, the circuit connection of the telephone system faces an electrical resistance across the line causing the current to flow through the wires connecting the telephone toward the central office. This current flows through a relay coil in a manually operated switchboard and actuates a buzzer or a lamp on the operator's switchboard to signal the operator for switching action.

7.3.2 Automatic Switching

Almon Strowger invented the automatic exchanges or dial service in 1888, and it had been used first commercially in 1892 [12, 13]. The very strong reason for the automatic switches was to eliminate the manual efforts while switchboard operations of fulfilling the required connections for a telephone call. This replacement of human operators at a telephone central office was done with an electromechanical system. These were electromechanical systems based on the components such as motors, shaft drives, rotating switches, and relays. Strowger switch or step-by-step switch, all relay, X-Y, panel switch, rotary system, and the crossbar switch were some of the types of automatic exchanges.

The telephone system was provided with a dial-in facility that helps the caller to transmit the destination telephone number to the automatic switching system. The mechanism of automatic exchanges was such that a telephone exchange will automatically sense an off-hook condition of the telephone whenever it was removed from on-hook or idle. A dial tone was provided by the exchange at that time indicating the user that the exchange is ready to receive user's dialed numbers. The pulses or dual-tone multiple frequency tones were generated from the telephone of the caller party and were to be processed to establish a connection with the called party within the same exchange or to another distant exchange. The connection established was to be maintained by the exchange until one of the parties hangs up the call.

A feature called the automatic number identification was implemented by the Bell System dial service to facilitate several of the services like automated billing, toll-free 800-numbers, and 9-1-1 service. In the manual switching exchanges, the operator was able to identify the origination of a call by seeing the light on the switchboard jack field. Long-distance calls were managed earlier by maintaining an operator queue before automatic number identification, where the operator asked and recorded the calling party's number on a paper toll ticket.

7.4 PACKET SWITCHING

Packet switching is such a technology that has a major contribution in revolutionizing the data communications [14–16]. Packet switching has got an important place in the networking history and is the basis for data communications in computer networks worldwide. This was the outcome of a continuous switching evolution in the telecommunication, specifically the time when computer data over the network was prioritized. All started with initially using the same circuit-switching technology that was used for voice communication in the telephone network, and virtually, all interactive data communication networks were circuit switched before 1968. The history of packet switching has its origin from the Advanced Research Projects Agency Network that we discussed in our earlier chapter, as the traditional voice-communication circuit switching was not found suitable for computer communication. The idea of switching with small blocks of data was conceptualized first by Paul Baran based on his research at the RAND Corporation in the early 1960s in the US. Donald Davies at the National Physical Laboratory had developed independently a similar message routing concept in the UK in 1965, and he was the one who coined the term packet switching and also proposed for a commercial nationwide data network in the UK. Lawrence Roberts, the director of the Advanced Research Projects Agency, was suggested with the Davies' proposal of packet switching at the 1967 Symposium on Operating Systems Principles and further used the packet-switching techniques invented by Davies and Baran in the development of the Advanced Research Projects Agency Network [17].

In the circuit switching, the network connection has to preallocate the transmission bandwidth for an entire call duration. However, it can be assumed that about 90% or more of this bandwidth is wasted because occurrence of interactive data traffic in short bursts. Thus, with advancement of technology and digital circuitry becoming cheap, it became more cost effective to think about the complete redesigning of data communication networks. While redesigning, the front runner approach was to introduce the concept of packet switching where the transmission bandwidth is to allocate dynamically. At the same time, it will also allow multiple users to share the same transmission line bandwidth that was earlier required for one user pairs. Packet switching has been so successful and has started not less than a revolution in the data communication industry. It gained so popularity over the span of time that in 1978, virtually all new developing data networks throughout the world were based on packet switching. The benefits of packet switching were limited to not only improve the economics of data communications but also enhance the reliability and functional flexibility of the network (Figure 7.4).

Packet switching has similarity with the message switching, where any message exceeding the defined maximum length in the network is broken down into shorter units, which is called as a packet. Messages in their entirety are routed in the message switching, which is store and forward of the entire message at a time. Taking inspiration from message switching's "store-and-forward" approach, packet switching must divide the complete message into packets, each with an associated header, that are independently transmitted through the network. The network links are getting shared by packets from multiple competing communication sessions instead of being dedicated to one communication session at a time in circuit switching. The header and payload are the components of a packet, such that the data in the header

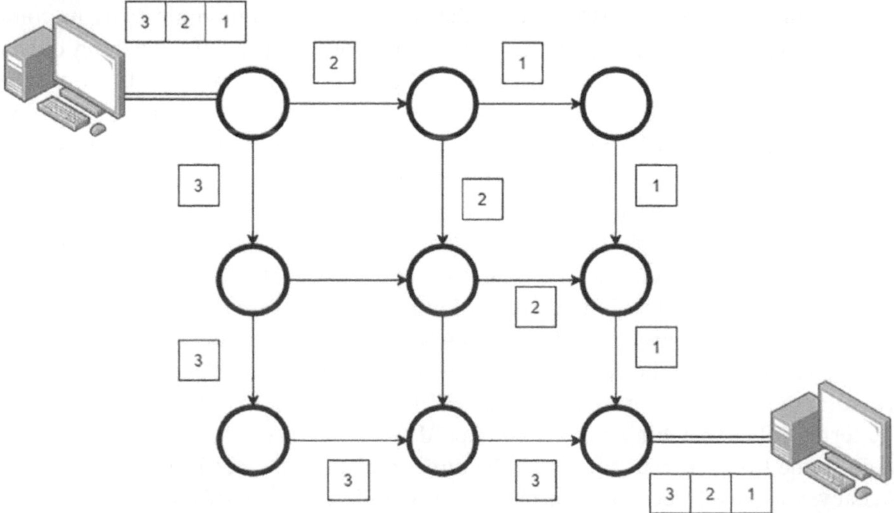

FIGURE 7.4 Packet-switching data network.

is used to guide the networking hardware to route the packet toward its destination in the network whereas the payload is the actual message's data that is to be extracted and used by application software [18].

The two general and popular techniques further used in packet switching are based on connection-oriented communication or connectionless communication, which is further based on virtual circuits (VCs) or datagrams.

7.4.1 Datagram

A datagram is a connectionless approach to packet switching where the data is divided into packets, which are also called as the datagram that is a basic transfer unit associated with a packet-switched network. The structure of a datagram is typically in header and payload sections and provides a connectionless communication service over a packet-switched network. The datagram delivery, arrival time, and order of arrival of datagrams are not guaranteed by the network in this data transmission scheme. The datagram scheme took inspiration from the telegram approach of communication, and the term "datagram" was created by combining the words data and telegram by the Consultative Committee for International Telephony and Telegraphy rapporteur on packet switching in the early 1970s. The datagram technique is compared often with a postal delivery system, where the sender has to only provide the destination address on the post or mail, without any guarantee of the delivery, and receives no confirmation upon successful delivery at the destination. Datagram services are unreliable, route datagrams without creating any predetermined path, and is therefore considered connectionless. The connectionless data transmission scheme is that method which is used for transmitting data through the Internet.

The datagrams are independently transmitted through the network, where each of the datagrams is labeled for ordering of the related packet with its destination and a sequence number. And it is not necessarily needing a dedicated path to transfer the datagram over the network; based on the header information, the packet can find its way to its destination. Each datagram has to be transmitted independently, with each taking a different route or transmitted via a different path in the network, and thus will not be received at the destination in the same order. Since datagrams arrive at the destination not in the same order in which they were sent by the source station, it is required that the datagrams should be numbered so as to reassemble them properly at the destination. The original message is obtained by reassembling of the received datagrams based on the packet sequence number to reproduce the original message at the destination. As a result, the datagram packet-switching technique allows multiple pairs of nodes to communicate concurrently over the same channel and does not require any kind of circuit to be established.

7.4.2 Virtual Circuit

VC is a method of data transmission over packet-switched network and uses the packet-switching technology that emulates a circuit switching. The transmission of the packet happens in such a way that it appears as though there is a dedicated physical link between the source and destination end nodes. The connection is to establish before transfer of any packet, and also, it provides ordered delivery of the packets. In the VC switching technology, the term "virtual circuit" is actually synonymous with virtual connection.

Since a VC is a connection-oriented transmission scheme, it requires that a connection to be virtually setup before the data transmission session should start, where each packet has to take the same route through the network. And therefore, all the packets of the data transmission session usually arrive in the same order at the destination in which they were sent by the source. Alternatively, in a connectionless scheme, each packet may take a different path through the network.

A virtual path is established first between the source and the destination nodes before the data communication takes place in VC – packet switching – and appears as if a dedicated physical path is present. But actually, in the network, it is a logical circuit association from a managed pool of circuit resources based on the requirements of network traffic. The network resources thus allow parts of this path to be shared by other communication sessions, that is however not visible to this user. Switching is done at the network layer of the communication system, and all the data packets transmitted will use the same path. In VC switching, all the network resources get reserved before the transmission like buffers and bandwidth, and all packets have to consume same resources. It requires a common header and use of routing information because all packets follow the same path. VCs provide a greater reliability and less complexity in the data transmission over the network, owing to fixed paths and fixed resources.

VC set up, data transfer, and teardown of the VC are the three phases of data transmission by the VC technique. A VC or a route is to be established first between the source and destination stations through number of switches in the first phase, i.e., setup of VC. In the data transfer phase, all the packets have to follow the same route established using the routing tables once the VC is set up. The teardown phase of the established VC starts with the source sending a teardown request and the destination responding with a teardown confirmation after the data transfer is complete. After that, the switches have to flush the entries in their routing table, helping in the tearing of the circuit.

7.5 CONCLUSION

We have gone through certain switching techniques in the history of telecommunication in this chapter which were responsible for establishing a network connection. Switches are the hardware devices that are used to connect two or more links and have been a very important and integral part of a network to extend the communication channel. Data moves in separate, small blocks known as packets based on the packet's destination address in a packet-switching network. These packets are further to be reassembled in the proper sequence to make up the message when received at the destination. However, the circuit switching requires dedicated point-to-point connections to the setup at the time of setting up a call. Both switching techniques in the networks have traditionally occupied different spaces within corporations. Circuit switching was used traditionally for making the telephone calls, and the packet switching was traditionally used to handle data.

REFERENCES

1. Maskara, S.L., Communication Networks and Switching, Nptel.ac.in. Accessed: May 20, 2021. [Online]. Available: https://nptel.ac.in/courses/117/105/117105076/
2. Virtual - Circuit Network, Tutorialspoint.com. Accessed: May 21, 2021. [Online]. Available: https://www.tutorialspoint.com/virtual-circuit-network
3. Wikipedia contributors, Circuit switching, Wikipedia. Accessed: May 24, 2021. [Online]. Available: https://en.wikipedia.org/w/index.php?title=Circuit_switching&oldid=1056948178
4. Brainerd, J. G. (1977). A History of Engineering and Science in the Bell System: The Early Years (1875–1925). Physics Today, 29(11), 64.
5. Joel, A. E. (1977). What is telecommunications circuit switching? *Proceedings of the IEEE*, 65(9), 1237–1253.
6. Scudder, F. J., & Reynolds, J. N. (1939). Crossbar dial telephone switching system. *The Bell System Technical Journal*, 18(1), 76–118.
7. RAND Corporation provides objective research services and public policy analysis, Rand.org. Accessed: May 30, 2021. [Online]. Available: https://www.rand.org/
8. Computer network, www.javatpoint.com. Accessed: May 25, 2021. [Online]. Available: https://www.javatpoint.com/computer-network-switching-techniques
9. TSSN tutorial, Tutorialspoint.com. Accessed: May 26, 2021. [Online]. Available: https://www.tutorialspoint.com/telecommunication_switching_systems_and_networks
10. Wikipediacontributors,Telephoneexchange,Wikipedia.Accessed:May27,2021.[Online]. Available: https://en.wikipedia.org/w/index.php?title=Telephone_exchange&oldid=1071666168
11. Switching systems in telecommunication networks, Jul. 5, 2018. Carritech Telecommunications. Accessed: May 28, 2021. [Online]. Available: https://www.carritech.com/news/switching-systems-in-telecommunication-networks/
12. Craft, E. B., Morehouse, L. F., & Charlesworth, H. P. (1923). Machine switching telephone system for large metropolitan areas. *Transactions of the American Institute of Electrical Engineers*, 42, 187–204.
13. Scowen, F. (1974). The automatic telephone exchange. *Transactions of the Newcomen Society*, 47(1), 35–46.
14. Wikipedia contributors, Packet switching, Wikipedia. Accessed: May 29, 2021. [Online]. Available: https://en.wikipedia.org/w/index.php?title=Packet_switching&oldid=1068604446
15. Wikipedia contributors, Datagram, Wikipedia. Accessed: May 29, 2021. [Online]. Available: https://en.wikipedia.org/w/index.php?title=Datagram&oldid=1060233958
16. Wikipedia contributors, Paul Baran, Wikipedia. Accessed: May 28, 2021. [Online]. Available: https://en.wikipedia.org/w/index.php?title=Paul_Baran&oldid=1062472769
17. Baran, P. (2002). The beginnings of packet switching: some underlying concepts. *IEEE Communications Magazine*, 40(7), 42–48.
18. Roberts, L. G. (1978). The evolution of packet switching. *Proceedings of the IEEE*, 66(-11), 1307–1313.

References

[illegible]

8 Multiprotocol Label Switching

ABBREVIATIONS

ARIS	Aggregate route-based IP switching
ARP	Address Resolution Protocol
ANSI	American National Standards Institute
ASIC	Application Specific Integrated Circuit
ATM	Asynchronous Transfer Mode
BISDN	Broadband Integrated Services Digital Network
DLCI	Data link connection identifier
EXP	Experimental Bit
FEC	Forwarding equivalence class
FIB	Forwarding information base
GSMP	General Switch Management Protocol
IETF	Internet Engineering Task Force
IFMP	Ipsilon Flow Management Protocol
ISP	Internet service provider
ITU-T	International Telecommunication Union - Telecommunication
LDP	Label Distribution Protocol
LER	Label edge router
LFIB	Label forwarding information base
LIB	Label information base
LSD	Label switch domain
LSP	Label-switched path
LSR	Label switch router
MPLS	Multiprotocol Label Switching
NP	Network Processor
OSI	Open Systems Interconnection
PPP	Point-to-Point Protocol
QoS	Quality of service
RFC	Request for Comment
RSVP	Resource Reservation Protocol
TDP	Tag Distribution Protocol
TFIB	Tag forwarding information base
TTL	Time to live
VC	Virtual channel
VCI	Virtual channel identifier
VP	Virtual path

DOI: 10.1201/9781003302902-8

VPI Virtual path identifier
VPN Virtual private network

8.1 INTRODUCTION

Multiprotocol Label Switching (MPLS) is an innovation in the packet-switching technology, used in the telecommunication networking. MPLS is a technology that has played an important role in making the telecommunication services cheap and of wide usage. Like other technological innovations, this too has a history and a beautiful journey of its implementation to getting popular in the telecommunication industry. At the same time, MPLS had redefined the service provider's networking and presented a cost-optimized network architecture at the core with a shared traffic approach. MPLS can be thought of as an innovation in the packet-forwarding technology and is a switching technology that regulates data traffic and packet forwarding in a complex network.

It has introduced a concept of labels as a differentiator for the data routing technique in a telecommunication network, directing data from one network node to the other based on short path labels rather than long network addresses. The label-based approach gives an edge in the packet-forwarding technique in telecommunication networks compared to the traditional preceding approaches in many ways and avoids complex lookups in a routing table and speeding traffic flows. Packet analysis happens at each hop, where the packet-forwarding decision is taken based on packet header analysis, followed by routing table lookup in the traditional Internet Protocol (IP) routing. On the other hand, in an MPLS network, labels are assigned to data packets, and the packet-forwarding decision is totally based on these label headers at each node. Thus, the MPLS technique has given a new approach of packet data handling, which is different from the conventional routing mechanism. In an MPLS network, a packet's header is analyzed only when it enters into the MPLS cloud, after which the packet-forwarding decision is "label-based" to ensure fast packet transmission between network nodes.

MPLS provides a connection-oriented methodology for fast packet transmission, traversing packets from the source to destination node across networks. Among its several important features, one of them is of encompassing packets in the presence of different network protocols.

8.2 MPLS BACKGROUND: A HISTORICAL VIEW

Before starting our discussion on the MPLS technology, let us have a historical view of the MPLS, from where it came and why it has been so popular. All of this started with telecom service providers/Internet service providers started deploying the Frame Relay, Asynchronous Transfer Mode (ATM), and IP-over-ATM in the 90s for the purpose of getting a technology that will be simpler, able to provide higher network throughput, and with lower latency. During the early stages of the evolution of service provider networking technology, looking for improvements was advantageous for a variety of reasons, including the fact that the devices and technology were not as advanced as they are today with high bandwidth links, and that a slight shift toward improvement was a welcome move. The demand for platform growth, network throughput, and faster packet handling was increasing with rapid popularization of the Internet in the early 1990s. But the hardware technology of supporting platform at that time was facing certain limitations; one of them was the packet lookup with longest address matching algorithm, and another was the hop-by-hop packet-forwarding mode. It was a major performance bottleneck in the technology that was limiting the packet-forwarding speed in the network. Thus, the search for a fast router technology was indeed a major research objective at that time.

ATM is a connection-oriented communication technology based on cell switching. It appeared first in the Broadband Integrated Services Digital Network and has been recognized as a useful technology, standards/specifications being developed by International Telecommunication Union – Telecommunication, American National Standards Institute, and ATM Forum. The concept of cell switching was able to provide a flexible and responsive method for multiplexing all forms of network traffic such as data, image, voice, and video, with all network traffic placed in fixed-length data packets called as cells to get switched at high speed. This technology received much attention due to its high data rate capacity, scalability of bandwidth, and the ability to support multiservice traffic in the ATM networks. ATM is, however, a connection-oriented communication standard, whereas the majority of modern day's data networking standards are supporting connectionless technology. This was the reason for applying ATM technology into data communication; this mismatch has led to complexity, inefficiency, and duplication of networking functionality.

The transport mechanism of ATM technology was based on virtual circuits, which were identified using identifiers such as virtual paths and virtual channels. To determine the next hop of the ATM cell, each of the ATM switches had to check the header of the cell containing virtual path identifier (VPI)/virtual channel identifier (VCI) values. An ATM switch was used to maintain a table with entries for every port/interface, for the forwarding decision of the ATM cell to take place. The forwarding table had entries with each row mentioning the condition to match VPI/VCI combination of the received ATM cell and then to rewrite it with some particular VPI/VCI combination before sending it out from a particular interface. Furthermore, to match the forwarding tables in adjacent ATM switches to point each other for a pair of communicating endpoints, a signaling control was established in ATM technology, such that, along the entire path, the outbound interface VPI/VCI combination of an upstream ATM switch would exactly match with an inbound interface VPI/VCI combination of the adjacent downstream switch.

In the IP-based technology, data forwarding was needing a router to receive the frame and perform a routing lookup, with viewing the IP header and matching the destination IP address with its routing table. The routing logic was to consider that entry from the routing table with longest possible prefix matching to the destination IP. The next logic of the routing lookup was that the router would need to perform another lookup for the next-hop IP address. It could potentially take several iterations to find a routing entry pointing out a specific interface with respect to the next-hop IP address. The logic next was of the router to know the

Layer 2 address with consulting the ARP table (or any other Layer 3 to Layer 2 mapping table) to be used while forwarding the packet through the immediate next hop. The router could send a packet out only after following all the process, also called as the process switching and was the basic mode of performing routing in IP technology.

ATM technology, at the same time, can provide faster performance than the IP routing because of the use of fixed-length labels and also maintains a much smaller label table than the comparative larger routing table. This further opened a debate at the time about which of the two data-forwarding technologies would be suitable in fast network implementation and considered as the basis of the next generation of network technologies.

IP packet technology was simple but limited in performance, whereas the ATM technology was able to provide high performance, having complex control signaling and high deployment costs. And the technology development started with some of the people trying to combine the advantages of both technologies, ATM and IP technology, for providing a simple and high performing technology. The result of that development has come up in the form of many probable solutions for label switching based on similar studies conducted by many vendors.

Frame Relay technology has the concept of data link connection identifiers in place of the VCI in ATM technology and was a locally significant value for each router. Data transport in the Frame Relay was possible by matching and rewriting these values between devices, without creation of any mapping, association, or changing of the frame format. Simplicity of the Frame Relay technology made it very efficient and attractive to the telecom service providers.

8.2.1 IP Switching Technology

Ipsilon Network, a Californian-based computer networking company, had launched "IP switching", a stunning technology called in the spring of 1996; they launched the first product known as IP Switch ATM 1600 in 1996. The technology they developed was based on combining the benefits of both the ATM technology and the IP technology. IP switching technology as ATM under IP was offered as a combination of simplicity, scalability, and robustness of the IP with the speed, capacity, and multiservice traffic capabilities of ATM [1].

The technology was proposed as an alternative technology discarding the end-to-end ATM connection and integrating fast ATM hardware directly with the IP, thus leveraging high-speed data-forwarding capability of ATM technology and preserving the connectionless nature of the IP. This was accomplished through the addition of an IP routing engine to ATM switches, resulting in a simple and easy deployment. It was using ATM hardware with soft-state to cache the IP forwarding decision, allowing the ATM hardware to switch the same IP flow traffic rather than forwarded by IP software. This had the benefit of not needing an end-to-end signaling, or address resolution, and requiring only the standard IP routing protocols. This approach directly supported the IP multicast and easily integrated into existing networks because it was based on IP. It would impose a heavy load on the ATM signaling protocol if this had to establish an end-to-end ATM connection for

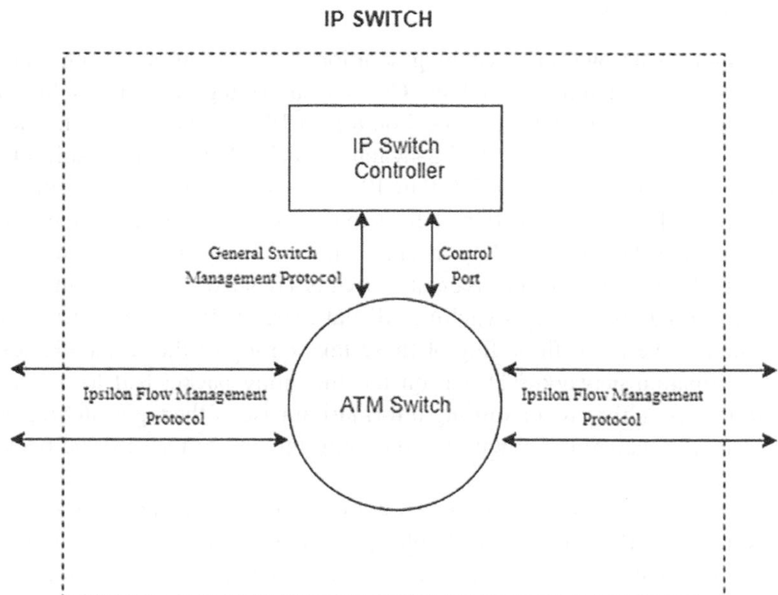

FIGURE 8.1 IP switch – structure.

every IP packet flow and would create unnecessary delay for query-response traffic. The structure of the IP switching developed by the Ipsilon Network is represented in Figure 8.1.

An IP routing engine has to run standard IP routing protocols and perform hop-by-hop IP forwarding in the IP switching, using the efficient forwarding plane of ATM and giving up the complicated control plane protocol of ATM. It also used the simpler protocol developed by Ipsilon including the Label Binding Protocol (Ipsilon Flow Management Protocol RFC1953) and Switch Management Protocol (General Switch Management Protocol RFC1987). These protocols were used to complete the ATM channels establishment between devices. A forwarding channel is established in the data-driven method in IP switching technology. The IP routing engine will negotiate with the upstream adjacency nodes when a service flow is detected through General Switch Management Protocol/Ipsilon Flow Management Protocol and has to assign a VPI/VCI for that service flow and then update the corresponding ATM switching table. This process was to repeat at each IP switching device on the routing channel of the service flow and to create an end-to-end ATM channel for the service flow. The service flow changes to the ATM forwarding mode from the hop-by-hop IP forwarding mode and can achieve higher forwarding performance in the network with this approach.

IP switching technology was a popular and impactful technology that led Ipsilon into a star of the IP communication industry. It had further inspired other giants of the communication industry like Cisco and IBM to go for a three-layer switching solutions and triggered a revolution of technology to lead the emergence of MPLS technology.

8.2.2 Tag Switching

Cisco announced its own tag switching solution, not long after Ipsilon Networks announced its IP switching technology. Cisco system's tag switching solution was based on the approach of putting a "tag" on top of IP packets, where a tag was the name that is now known as a label. Tag switching technology got its first implementation release in Cisco IOS 11.1(17)CT in 1998. This implementation could assign tags to networks for each destination address in the routing table, with the purpose of establishing a tag (label) forwarding mechanism in the network. Furthermore, a tag forwarding information base table is created, that is a table to store the input-to-output tag/label mappings in the tag switching. IP data stream from the routing table is encapsulated in the tag with putting of those tags on top of the IP packet destined for that destination network. The tag on the incoming packet had to match each time with the tag in the tag forwarding information base of the tag switching router, then to swap the incoming tag with the outgoing tag and to forward the packet [2] (Figure 8.2).

Cisco System's "tag switching" technology is quite different from that of the Ipsilon Network's "IP switching" technology, such that a forwarding channel is created in the tag switching with respect to each destination address listed in the tag switching router's routing table. On the contrary, the forwarding channel is driven by the data flow and is used only for the data flow in the IP switching technology.

8.2.3 IBM ARIS

After the announcement from Cisco for its tag switching technology, there was an announcement from IBM on a proposed tag switching solution called aggregate route-based IP switching (ARIS). IBM's proposal of ARIS has much in common

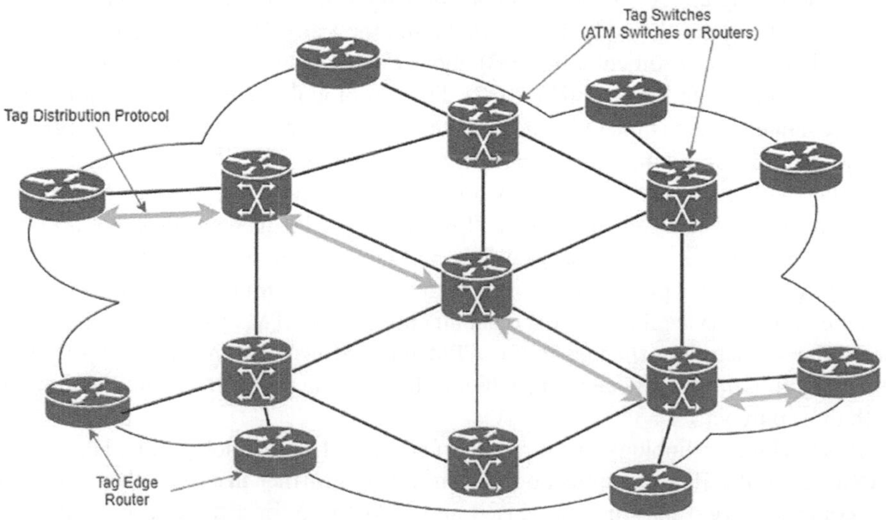

FIGURE-8.2 Tag switching architecture.

with tag switching, and the ARIS technology had similarity with Cisco's tag switching. Both of switching technologies were based on topology-driven tagging technologies, especially considering the handling of destination-based routing. However, ARIS considered initially the ATM network as the link layer and proposed the virtual circuit merging technology to eventually integrate into MPLS. ARIS technology takes the advantage of integrated router switches with mapping the routing information to short fixed-length labels. These labels then help to easily determine the packet's next-hop router by direct indexing, rather than using the router's standard packet evaluation and routing lookup process. Based on the forwarding paths that are already established by router's routing protocols, ARIS set up a virtual circuit throughout a network [2].

Both of switching technologies, the "tag switching" and "ARIS", were invented in parallel; the proposal of ARIS was announced following the proposal of tag switching by only a few weeks. Proposals to both technologies rely on some previously published ideas, such as the threaded indices notion as was described by Varghese in 1995. Unlike Ipsilon's IP switching, the naming convention of ARIS as "aggregate route-based IP switching" mentions something about its origins, such as the binding of labels to aggregate routes (or pool of network addresses) in ARIS, rather than to flows in Ipsilon's IP switching. And therefore, like IP switching, ARIS is not a flow-driven scheme, but it is a tagging or labeling scheme like MPLS and Cisco's tag switching.

8.2.4 MPLS

We have gone through some of described vendor-specific approaches to multilayer routing as mentioned above that had appeared between 1994 and 1997. The mechanism of multilayer routing has to cover the approaches for integration of Layer 3 packet forwarding and Layer 2 switching. The label-based lookups were preferred over the hop-by-hop routing lookups for more efficient packet transmission, along with the potential to engineer the network and manage the impact of data flows. MPLS working group of the Internet Engineering Task Force was formed in 1997, and the MPLS working group released its first MPLS RFCs in 2001, where MPLS architecture was specified in RFC 3031 and MPLS label stack encoding was specified in RFC 3032. The formation of the MPLS working group of the Internet Engineering Task Force was a need of the hour due to the fact that these approaches that we just discussed above were proprietary and produced incompatible network solutions [3].

The label-switching feature of MPLS allows the switching operations to a device in the form of labeled – IP packets with performance similar to an ATM switch. Like the ATM switches the label lookups in MPLS is also faster comparatively to a conventional IP routing. MPLS overcomes ATM setbacks with the advancement in the packet switching by providing less packet overhead and connection-oriented services for frames with varying length. At the same time, the MPLS provides the advantages of maintaining traffic engineering [4, 5] and out-of-band control. With the evolution of the MPLS, the Frame Relay and ATM network technologies have become less in need for installing large-scale networks because the performance of MPLS has been far superior to that of the previous switching technologies [1].

MPLS technology uses the method of encapsulation containing network layer grouping with labels and can obtain link layer services from various link layers (e.g., Point-to-Point Protocol, ATM, Frame Relay, Ethernet) to provide connection-oriented services for the network layer. MPLS proposal was originally a protocol for the purpose of improving the forwarding speed of routers in the network. However, with the advancement in the hardware technology, high-speed routers and widely used three-layer switching architecture adopting ASIC and NP forwarding, the original MPLS intention of improving the forwarding speed has become meaningless. The IP routing protocol and control protocol can support MPLS and can be supported from powerful and flexible scheme of policy-based constrained routing to meet the network requirements of various new applications. Originating from the IPv4, the MPLS technology can be extended to support many network protocols (IPv6, IPX, and so on).

Amidst the advantages of MPLS, there are still few downsides as well for the label-switching technology. One among them is the complicated management of MPLS with its dependence on the routing protocols for data transfer, such that the whole transportation in the network can be disturbed and will need redirection of data packets for any crumples in the network.

Multiple tags support, connection-oriented characteristic, virtual private network (VPN), traffic engineering functionality, quality of service, expansibility, and unification of MPLS/IP network architecture to provide clients with all kinds of service possible have made the MPLS technology an increasingly basic technology of large-scale network.

8.3 ARCHITECTURE OF MPLS

MPLS, as we have discussed, is a label-switching technology that makes forwarding decisions based on a label, so the concept of label is very important in the switching [6–14]. A label is a small and fixed-length entity, but it is not like an IP address or any other packet header information. The value of the label changes as the packet travels from one node to another node in the network. We are going to discuss about the architecture of MPLS and how the data flow in this section. Most of the concepts standardized in MPLS architecture are inherited from the tag switching by the Internet Engineering Task Force MPLS working group.

MPLS is technically neither a network layer or Layer 3 protocol nor a data link layer or Layer 2 protocol; it technically sits between both layers, so it is sometimes referred as a Layer 2.5 protocol. Being a layer 2.5 protocol gives MPLS flexibility of use with both Layer 3 and Layer 2 and gives MPLS the speed of Layer 2 and dynamics of Layer 3 technologies. Instead of Layer 3 addressing using IP addresses and Layer 2 addressing using the MAC addresses, the inserted labels are residing between Layer 2 and Layer 3 of OSI to take the forwarding decisions of the data packets based on the labels. This reduces the overhead from the conventional IP lookup in the complex IP tables and makes more efficient forwarding decisions. While having a VPN implementation using MPLS, we can set up the MPLS Layer 3 VPN with labels based on IP addresses, and we can also set up Layer 2 VPN using labels based on MAC address [15] (Figure 8.3).

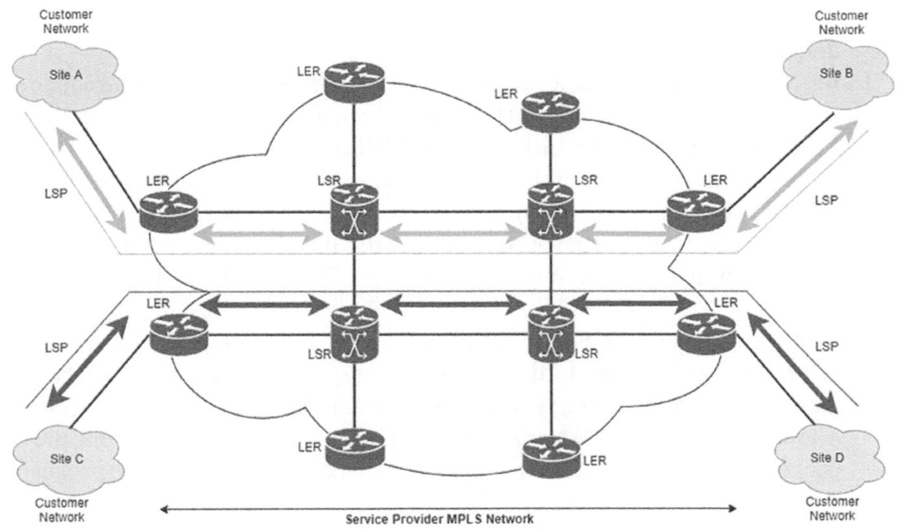

FIGURE 8.3 MPLS network architecture.

FIGURE 8.4 Structure of the MPLS label header.

8.3.1 MPLS Terminologies

We are going to discuss about some terminologies that are commonly used in MPLS, considering the MPLS network diagram represented in Figure 8.4.

Label Switch Domain: A label switch domain (LSD) is the MPLS network domain where the network devices will support the forwarding of traffic using MPLS technology. It is defined by the network administrator and contains one or more physical networks.

Label Edge Router: A label edge router (LER) is that router which is placed at the boundary of the MPLS network or the LSD. The LER is used to connect the customer network on one end with the service provider's MPLS network.

The traffic destined to and originating from the customer network will always be a nonlabel packet between the customer network and the LER, whereas the packet within the MPLS domain or LSD will always be encapsulated into a label. Therefore, the LER has very important responsibilities; one of them is of removing the label from a labeled packet whenever the packet has to cross the MPLS domain and to reach the customer network and another responsibility is of attaching a packet (nonlabeled) when an

LER receives it from the customer network before sending it into the MPLS domain.

Label Switch Router: A label switch router (LSR), also called as a provider router or transit router, is that router which is within the LSD or the MPLS domain after the LER. It is an MPLS router that has to perform switching action based on the label, and therefore, the data packet which it will receive or perform action will always contain the label. This type of router has to always be present in the middle or core of an MPLS network and will have to be surrounded by the LER at the edge of the MPLS domain.

After receiving a packet, an LSR has to use its label present in the packet's header as an index to determine the next hop on the label-switched path (LSP) and, accordingly, the corresponding label for the packet from a lookup table. After that, the LSR then removes the old label from the packet's header and replaces it with the new label before sending the packet containing new label to the next hop.

Label-Switched Path: A LSP is always established by the MPLS network provider before starting the packet transmission containing labels for many reasons, such as to create MPLS-VPN to route traffic along specified paths through the MPLS network. Therefore, it can be said that an LSP is a dedicated logical path that will be used by packets containing the label to carry the traffic from one fixed node to another fixed node, where the LSP will be connecting both of these nodes at its end. These LSPs in the MPLS are not different from the permanent virtual circuit in ATM or Frame Relay networks in many respects, except that they are not dependent on a particular Layer 2 technology.

Label Distribution Protocol: A Label Distribution Protocol (LDP) is one of such protocols that are used to distribute the labels within the LSD to form the LSP. Protocols like the LDP or Resource Reservation Protocol may be used for distribution of labels between LERs and LSRs.

In the MPLS network, LSRs have to regularly exchange the label and reachability information with each other. This exchange of information is done based on standardized procedures to create a complete picture of the network connectivity to use that information while forwarding packets.

Forwarding Equivalence Class: A forwarding equivalence class (FEC) is a group or flow of packets to get forwarded along the same path with the same forwarding treatment, and thus, the packets belonging to the same FEC should have the same label. However, on the other hand, not all packets containing the same label belong to the same FEC, due to the reason that their experimental bits (EXP) values may differ; can have different forwarding treatment; and could belong to a different FEC. The ingress LSR is the router to determine which of the packets belong to which FEC; has to always classify packets received and to attach labels to the packets; and, therefore, is responsible for determining the FEC to which a packet belongs.

8.3.2 MPLS Label Header

A label consists of a 20-bit identifier to uniquely identify an FEC belonging to an IP packet. After receiving an IP packet from the end customer or a non-MPLS network, the LER or edge LSP has to create an MPLS header in the packet and insert a specific label as a packet's index into this field. This labeled IP packet is then to be considered as an MPLS packet to traverse through the MPLS domain with label-based operation. The label assigned is meaningful only to a local end. Figure 8.4 represents a 4-byte MPLS label header structure.

The following are the fields contained in an MPLS header:

- **Label:** It is a 20-bit field identifier that identifies a label value.
- **EXP:** It is a 3-bit field mentioned as EXP or Experimental bits which is used for extension; the three EXP are used by the class of service function for quality of service and is similar to Ethernet 802.1p.
- **BoS:** It is a 1-bit field that identifies the bottom of a label stack; hence, it is called as the bottom of stack bit. In the MPLS network, a packet can be supported with multiple labels, which are stacked to form a label stack. This 1-bit bottom of stack is the identifier, such that if the bottom of stack field of a label is set to 1, then it will mention that the label is at the bottom of the label stack.
- **TTL:** It is an 8-bit field to indicate the time to live TTL value and will do the same function as the time to live field in IP packets.

Encapsulation of labels is done between the data link layer (Layer 2) and the IP/network layer (Layer 3) and supported by all the protocols of the data link layer, as illustrated in Figure 8.5, mentioning the position of the label in a packet.

We have discussed that an MPLS packet can contain stack of labels that may have an ordered set of labels. Figure 8.6 illustrates a label stack, where the label near to the Layer 2 header is called the stack top label or outer label. On the other hand, in the label stack as mentioned, the label near the Layer 3 header is called the stack bottom label or inner label.

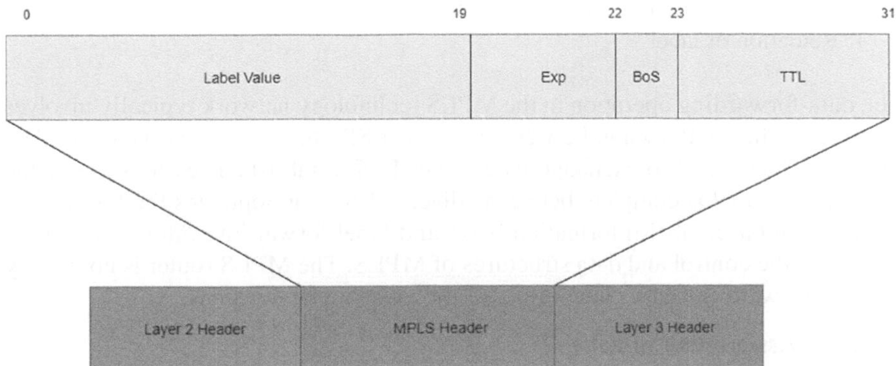

FIGURE 8.5 MPLS label position in a packet.

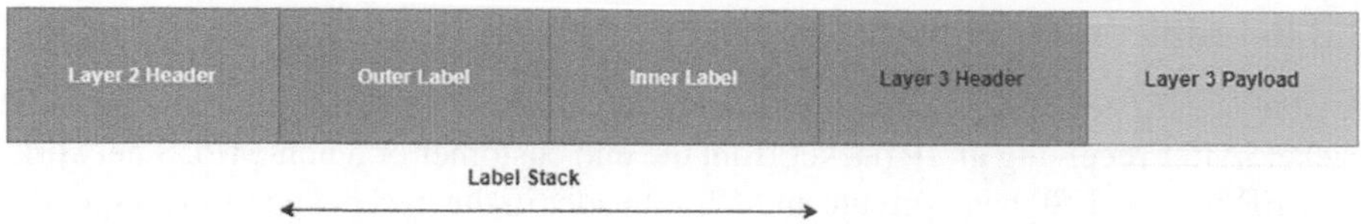

FIGURE 8.6 MPLS label stack.

8.3.3 MPLS Operations

The MPLS starts with an establishment of the LSPs, which are predetermined paths and unidirectional between pairs of routers across an MPLS network seeking the end-to-end delivery of the packet, when a packet from the customer network enters the MPLS network domain through a LER or the edge LSR (also known as an "ingress node"). The received packet is then assigned to a FEC by the LER, which will depend on the type of packet data and its intended destination, where identification of packets with similar or identical characteristics is done by the FEC. The LER will apply a label to the packet based on the FEC and encapsulate the labeled packet inside an LSP.

The packet is supposed to traverse through and follow the assigned LSP to be received at its intended destination. The LSP consists of transit nodes called as the LSRs as the labeled packet moves through the network, and those LSRs have to direct the data by the instructions in the packet label. The actions performed in between these LSRs are based on the packet label, without any additional IP lookups, and thus, the process is called as label switching. Labels on the packet can also be stacked so that multiple labels are present on the IP packet. In that case, the top label controls delivery of the packet. The top label is "popped" and the label underneath takes over for direction when it reaches its destination.

The label is removed from the packet to deliver the IP packet via the IP routing mechanism at the "egress node", or destination LER (final router at the end of the LSP).

In the MPLS technology, forwarding of data involves the following four steps:

1. Assignment of label
2. LDP or Tag Distribution Protocol (TDP) session establishment
3. Distribution of label (using a label distribution protocol)
4. Retention of label

The data-forwarding operation in the MPLS technology network typically involves forming of the LDP session between adjacent LSRs to assign local labels to destination prefixes and to exchange labels over LDP established sessions. After the exchange of label is complete between adjacent LSRs, it populates the forwarding information base, label information base, and label forwarding information base, which are the control and data structures of MPLS. The MPLS router is now ready and can forward the data plane information based on label values.

8.3.3.1 Assignment of Label

A label has to be assigned first to a router's reachable IP networks, and then, these labels are to be imposed on the forwarded data packets to those IP networks.

Destination networks reachability information is advertised by the IP routing protocols; MPLS technology will use the similar process for routers or devices of the MPLS domain to learn the labels assigned to destination networks by neighboring routers. The LDP or TDP has to assign labels and exchange them following session establishment between adjacent LSRs in an MPLS domain.

8.3.3.2 LDP or TDP Session Establishment

Labels are to be distributed after the label assignment on a router is done among directly connected LSRs if the MPLS forwarding has been enabled on the connecting interfaces between them. Two very popular protocols which are commonly used for the distribution of labels are the LDP or the TDP. The TDP was the earlier protocol we discussed in the previous section which is now deprecated, and thus the LDP is the label distribution protocol by default for many of device interfaces, which is to be enabled at the time of router configuration. Both protocols, LDP and TDP, generally function the same way but are not interoperable, and both the TDP and LDP can be used by a router on the same interface for enabling dynamic formation of the LDP or TDP peers as per the protocol running on the interface of the peering MPLS neighbor (Table 8.1).

There are four categories of LDP messages for session establishment.

8.3.3.3 Distribution of Label

A label has to be assigned to an IP packet based on the destination prefix and the information is updated in the router's forwarding information base in an MPLS domain running LDP. The label is then to be distributed toward the upstream neighbors in the MPLS domain after the session is established. The router's local significant labels are to get exchanged with adjacent LSRs during the process of label distribution. It will then be followed by storing of specific prefix to a local label and a next-hop label in the label forwarding information base and label information base tables in the routers, for the purpose of label binding as per information received from the downstream LSR.

The label distribution modes working in MPLS are mentioned below:

- **Downstream on Demand:** This mode of label distribution is based on allowing an LSR to explicitly request from its downstream next-hop router for a mapped label to a particular destination prefix. Therefore, this technique of label distribution is called as the downstream on demand label distribution.

TABLE 8.1
LDP Session Establishment Messages

Message's Detail		
1	Discovery messages	Announce and sustain an LSR's presence in the network
2	Session messages	Establish, upkeep, and tear down sessions between LSRs
3	Advertisement messages	Advertise label mappings to FECs
4	Notification messages	Signal errors

- **Unsolicited Downstream:** This mode of label distribution is based on allowing an LSR to distribute bindings to upstream LSRs without waiting for a request explicitly coming from them. Therefore, this technique of label distribution is called as the unsolicited downstream label distribution.

8.3.3.4 Retention of Label

LSRs supporting "liberal label retention mode" have to maintain the label bindings with destination prefix, for those labels received from the downstream LSRs which may not be the next hop for that destination. However, the LSRs supporting a "conservative label retention mode" have to discard the label bindings received from those downstream LSRs, which are not the next-hop routers to that destination prefix.

Therefore, with a liberal retention mode, an LSR can almost immediately start forwarding labeled packets and is more memory consuming due to the reason that the numbers of labels maintained for a particular destination are large. Whereas, on the other hand, with the conservative label retention mode is minimal memory consuming as the labels to be maintained are labels from the confirmed LDP or TDP next-hop neighbors.

8.3.4 Label Operations

We discussed about the MPLS operation; the MPLS technology has the data-forwarding method using the label switching, and thus, label operation is very important to understand MPLS technology. All the LSRs along an LSP in an MPLS domain have to examine the MPLS label, have to determine the LSP next hop, and have to then perform the required label operations.

There are five label operations that an LSR can perform:

Push Operation: During the push operation, an LSR has to add a new label on top of the packet. When an IPv4 packet is arrived at the LER (or edge LSR) inbound to the MPLS domain, the LER has to add a label to the IP packet, and the added new label is the first label in the label stack. With this operation, a label is added to the stack, and it has to set the stacking bit (bottom of stack bit) to 1 or high, to indicate that this is the bottom of stack. However, if the MPLS packet already has an existing label, with push operation, more labels that will set the bottom of stack bit to 0 are added to indicate that more MPLS labels have to follow the first label or the current label (top label) in the label stack is not the bottom label.

When an LER (edge LSR) receives the IP packet, then the LER has to perform an IP route lookup for the packet to identify the label for the destination prefix. With the route lookup an LSP next hop is identified and the LER has to perform a label push operation on the packet with adding suitable label to the packet, and then to forward the labeled packet to the LSP next hop.

Swap Operation: During a label swap operation, an LSR has to replace the top label of the label stack with a new label. An LSR has to perform an MPLS forwarding table lookup when a transit router receives the packet. The label lookup in the label forwarding information base table provides the LSP next hop and the corresponding label of the link between the transit LSR and the next-hop router in the LSP. Accordingly, the LSR replaces the top label of the label stack as per the next-hop label and sends the packet to the next hop.

Pop Operation: During the pop operation, an LSR has to remove the top label from the label stack. The label pop operation performed at the penultimate router has to remove the entire MPLS label from the label stack. At the same time, this operation removes only the top label from the label stack for MPLS packets with an existing label and modifies the stacking bit as necessary (for example, sets it to 1 if only a single label remains in the stack).

Penultimate Hop Popping: At the last hop in the LSP of an MPLS network or the exit/egress LER, the label on the packet becomes unnecessary. Thus, a process can be used by which the penultimate LSR of the exit LER will be enabled to remove a label from a packet to be sent to the egress network (customer network or non-MPLS network). Many LSPs may share the same egress in the real-world situations of the MPLS network implementation, where with the process of penultimate hop popping, the burden on the egress router is reduced. The egress router is to forward the packet over an IP route to the final destination network. On the penultimate LSR, the configuration of penultimate hop popping is done to assign Label 3 on the egress. This Label 3 assigned indicates that an implicit-null label is present which never appears in a label stack. Once finding an implicit-null label or Label 3, the receiving router or edge LSR will directly remove the existing label, and the LSR does not need replacing of the existing label.

Multiple Push Operation: With this label operation, an LSR can add multiple labels to the top of the label stack, and thus, this action is equivalent of multiple push operations.

Swap and Push Operation: This label operation has to replace the top label with a new label and then again has to push a new label to the top of the label stack.

8.4 CONCLUSION

MPLS is a protocol-agnostic data-forwarding technique that has several benefits along with increase in the data transfer speed, quality of service, and traffic engineering, and it controls the flow of network traffic. It has played an important role in making the telecommunication services cheap and of wide usage, thus helping the service provider networks as well as the enterprise networks. It has redefined the service provider's networking with a cost-optimized architecture at the core in a shared traffic model and can be thought of as an innovation in the packet-forwarding technology. MPLS is a switching technology that regulates data traffic and packet forwarding in a complex network such that the data is directed through a label-switched path via labels instead of complex routing lookups in a routing table at each hop. MPLS works with both IP and ATM and is considered as a scalable and protocol-independent technology.

REFERENCES

1. Newman, P., Minshall, G., & Lyon, T. L. (1998). IP switching-ATM under IP. *IEEE/ACM Transactions on Networking*, 6(2), 117–129.
2. Agrawal, S. (1997). IP Switching. Online&excl Retrieved from Internet: Aug, 16. [Online]. Available: https://www.cs.wustl.edu/~jain/cis788-97/ftp/ip_switching.pdf
3. De Ghein, L. (2016). *MPLS Fundamentals: MPLS Fundamentals ePub_1*, Indianapolis, IN: Cisco Press.
4. Awduche, D. O. (1999). MPLS and traffic engineering in IP networks. *IEEE Communications Magazine*, 37(12), 42–47.
5. Xiao, X., Hannan, A., Bailey, B., & Ni, L. M. (2000). Traffic engineering with MPLS in the Internet. *IEEE Network*, 14(2), 28–33.
6. Armitage, G. (2000). MPLS: The magic behind the myths [multiprotocol label switching]. *IEEE Communications Magazine*, 38(1), 124–131.
7. Davie, B. S., & Rekhter, Y. (2000). *MPLS: Technology and Applications*, Burlington, MA: Morgan Kaufmann Publishers Inc.
8. Alwayn, V. (2001). *Advanced MPLS Design and Implementation*, Indianapolis, IN: Cisco Press.
9. Cisco learning network, Cisco.com. Accessed: June 2, 2021. [Online]. Available: https://learningnetwork.cisco.com/s/article/MPLS-History-and-building-blocks
10. Wikipediacontributors, MultiprotocolLabelSwitching, Wikipedia, Accessed: June3, 2021. [Online]. Available: https://en.wikipedia.org/w/index.php?title=Multiprotocol_Label_Switching&oldid=1070865875
11. Farrel, A. (2004). Multiprotocol label switching (MPLS). *The Internet and Its Protocols*. Elsevier, 385–490.
12. History of MPLS, Scribd. Accessed: June 5, 2021. [Online]. Available: https://www.scribd.com/document/95829703/History-of-MPLS
13. MPLS Fundamentals, Oreilly.com. Accessed: June 7, 2021. [Online]. Available: https://www.oreilly.com/library/view/mpls-fundamentals
14. M. Tripathi, Multiprotocol label switching(MPLS) explained, Towards Data Science, Aug. 10, 2019. Accessed: June 10, 2021. [Online]. Available: https://towardsdatascience.com/multiprotocol-label-switching-mpls-explained-aac04f3c6e94
15. Pepelnjak, I., & Guichard, J. (2002). *MPLS and VPN Architectures* (Vol. 1), Indianapolis, IN: Cisco Press.

9 Metro Ethernet

ABBREVIATIONS

API	Applications program interface
CSMA/CD	Carrier-sense multiple access with collision detection
CEN	Carrier Ethernet networks
ENNI	External network-to-network interface
EPL	Ethernet private line
EP-LAN	Ethernet private LAN
EVC	Ethernet virtual connection
EVPL	Ethernet virtual private line
EVP-LAN	Ethernet virtual private LAN
FR	Frame Relay
IP	Internet Protocol
LAN	Local area network
L3 VPN	Layer 3 virtual private network
LSO	Lifecycle Services Orchestration
MAC	Media access control
MAN	Metropolitan area network
MEF	Metro Ethernet Forum
MEN	Metro Ethernet network
MPLS	Multiprotocol Label Switching
OAM	Operations and maintenance
OPEX	Operational expenses
OSI	Open System Interconnection
OVC	Operator virtual connection
P2MP	Point-to-multipoint
P2P	Point-to-point
PVC	Permanent virtual circuits
SDH	Synchronous digital hierarchy
SD-WAN	Software-defined wide area networking
SONET	Synchronous optical networking
TDM	Time-division multiplexing
UNI	User–network interface
VPLS	Virtual private LAN service

9.1 INTRODUCTION

Ethernet has covered a long journey from its inception and has achieved many milestones since then. Modern-age Ethernet is completely different from the original Ethernet conceived by researchers, including Robert Metcalfe and David Boggs, at

DOI: 10.1201/9781003302902-9

Xerox PARC in the early 1970s. Early Ethernet that we have gone through in detail in Chapter 4 was a single-wired media called as "Ether"-based connection to share with hundreds of computers. The original Ethernet technology involved the key concept of avoiding collisions since data could not be received properly if multiple computers are transmitting simultaneously. Since its invention, the technology and architecture of Ethernet have gone through many changes. We have gone through the general Ethernet concepts in Chapter 4, which include architecture of Ethernet with respect to the Open System Interconnection reference model, MAC, Ethernet frame format, Ethernet addressing schemes, the carrier-sense multiple access with collision detection protocol, and the concept of repeaters.

In this chapter, we are going to learn about the concepts of metro Ethernet technology, a technology that is very popular in extending the service provider's network to give Ethernet-like feature. Early technologies were limited to a single link known as local area networks (LANs). From a multiple access protocol in a LAN, advancement in the Ethernet technology has placed it as a ubiquitous transport technology reaching into metro and wide area networks (WANs). It has been the indisputable and the most preferable technology of choice for LAN connectivity for nearly 40 years because of its several advantages. This chapter is going to provide us more information on understanding the concepts and underlying technologies for extending the Ethernet services, i.e., from basic Ethernet to carrier-grade Ethernets. The metro Ethernet network is one such effort of establishing or extending an Ethernet LAN technology to the metropolitan area networks (MANs), through the service provider's network. The metro Ethernet network has to replace the traditional voice-grade leased line service with legacy technologies based on time-division multiplexing and the synchronous optical networking/synchronous digital hierarchy due to the fact that these legacy technologies are inadequate for handling the high bandwidth demands of emerging data applications.

9.2 METRO ETHERNET

A MAN or a metro network has always been a challenging environment to deliver data services due to the reason that a metro network was built to handle the stringent reliability and availability needs of voice communications. That has further led to numerous evolutions in the metro networks worldwide, depending on many factors. A metro Ethernet is one such evolution of the metro networks, with not necessarily all Ethernet Layer 2 networks, but a deployment in combination with other technologies such as next-generation synchronous optical networking/synchronous digital hierarchy and Internet Protocol (IP)/Multiprotocol Label Switching (MPLS) networks [1].

Metro Ethernet is used by the service providers to provide the Layer 2 Ethernet connections through service provider's network in MANs between customer sites. We have already gone through the advantages of the Ethernet technology, which are its relative simplicity, high bandwidth, and low-cost switches that have further driven Ethernet to become the transport technology of choice in MANs as well. Modern day's network applications in the MAN include numerous applications requiring pure Layer 2 connectivity for providing simple point-to-point (P2P), point-to-multipoint

(P2MP), or multipoint-to-multipoint connectivity services at relatively low number of customer sites.

The limitations of the Ethernet however become apparent in large MANs with thousands of access nodes, and thus, it was more likely for the service providers to offer an MPLS Layer 3 virtual private network (L3 VPN) services. Due to this reason, the MPLS L3 VPN–based approach gives more flexibility, better scale, and ease of operations and maintenance to interconnect hundreds or thousands of customer sites. Consider, for example, the end-to-end Layer 3–based LTE mobile backhaul networks connectivity provided in the MAN by means of L3 VPN services. However, the L3 VPN–based solution is generally coming at higher cost per port in comparison to Ethernet switches–based Layer 2 services but has low operational expenses due to ease in network operations.

Ethernet technology was originally not designed considering the metropolitan networks or metro applications, and the Ethernet technology was lacking in the requirement of scalability and reliability for mass deployments. These parameters of scalability and robustness present in the IP and MPLS control planes were also required for deploying Ethernet in the MANs. A typical solution that emerged was a hybrid approach, where the Layer 2 technology and Layer 3 IP and MPLS have come together to give Ethernet's simplicity and cost effectiveness with scalability of IP and MPLS networks. The mixed Layer 2 and Layer 3 services complement each other in the modern MAN.

9.2.1 Metro Ethernet Forum

The Metro Ethernet Forum (MEF) [2–8] was established as a consortium of non-profit international industry in 2001, which was dedicated originally to Carrier Ethernet networks (CENs) and services. It was started as a group of smart, passionate, and dedicated people that has members from some partners, some suppliers, and some competitors coming together with the purpose of solving an industry problem. And that industry problem was about expanding "Ethernet" rapidly, a popular implementation of a LAN technology into the WAN and that the WAN can be able to achieve the Ethernet-like capabilities. The efforts of the people of this group helped in transforming this great technology of Ethernet (LAN technology) into a viable, business-class, implementation-agnostic connectivity service standard as a carrier-grade Ethernet service. MEF has empowered service providers in delivering standards-based Carrier Ethernet services, with assuring high-speed and lower cost connectivity services that can span multiple provider domains easily based on the integrity of industry certification. This group was formed as MEF, which created the globally adopted Carrier Ethernet Standards [9].

Later, MEF expanded its scope of work starting from 2015 and included additional underlay and overlay services as well. MEF included optical transport and IP as the underlay connectivity services and also includes overlay services such as the software-defined WAN.

Further to this came the digital transformation, and then MEF quickly and resolutely broadened its scope of addressing the need of modern enterprise's toward achieving more agile, dynamic, and automated connectivity services, digital services, and beyond. And thus, MEF also included works for orchestration and automation with its Lifecycle Services Orchestration framework and related interface reference points and applications program interfaces.

MEF is now a global industry forum for network and cloud providers, with a mission of developing a global federation of network, cloud, and technology providers to establish dynamic, assured, and certified services empowering enterprise digital transformation. The MEF 3.0 Global Services Framework is based on four pillars as follows: Services, Lifecycle Services Orchestration Applications Program Interfaces, Community, and Certification.

9.2.2 MEF: Carrier Ethernet Terminologies

As specified by the MEF, the following are the important terminologies which we will discuss [10] (Figure 9.1).

User–Network Interface: A user–network interface (UNI) is mentioned as the physical port or interface, which is the point of demarcation between the service provider or carrier's network and the customer. The service provider is to provide always the UNI, which is a standard physical Ethernet interface in the Carrier Ethernet network that generally has the port speed of 10 Mbps, 100 Mbps, 1 Gbps, or 10 Gbps.

External Network-to-Network Interface: An external network-to-network interface, as mentioned by the MEF, is the physical Ethernet port used to interconnect two Ethernet MANs of two different service providers on the service provider's access node.

Ethernet Virtual Connection: An Ethernet virtual connection (EVC) represents the architectural construct used for establishing Ethernet flows connection with the UNI reference points on the subscriber sites across the MAN.

The EVC is considered as an important attribute to the metro Ethernet service and has been defined by the MEF as "an association of two or more UNIs". There are two functions that an EVC performs:

1. An EVC facilitates in connecting two or more UNIs to enable transfer of Ethernet frames between them.
2. An EVC also facilitates in preventing the transfer of data from sites which are not part of the EVC, or EVC provides isolation of traffic between two UNIs not belonging to that EVC.

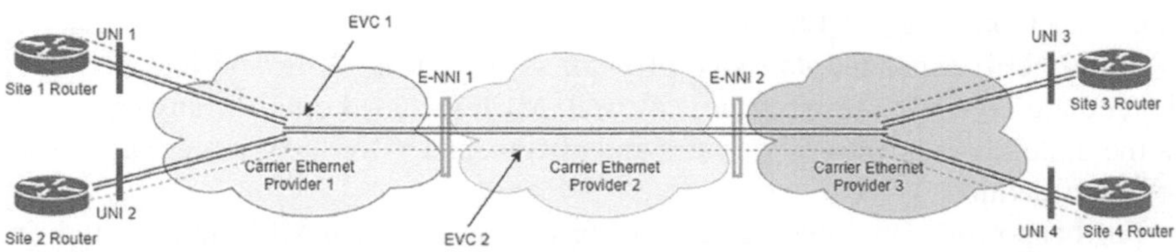

FIGURE 9.1 Network architecture of Metro Ethernet.

For any of the service type defined under the metro Ethernet, EVC forms a basis of defining it that depends on the number of UNIs (customer sites) and the method of connecting them. The MEF has defined three types of EVCs: the E-Line, E-LAN, and E-Tree.

9.2.3 MEF: Carrier Ethernet Services

As defined by the MEF, the below-mentioned diagram, Figure 9.2, represents different service types of the Carrier Ethernet that we are going to discuss in this section.

Considering the diagram above, we are seeing that each of the Carrier Ethernet service is available in two variants, which are port based and virtual LAN (VLAN) based.

9.2.3.1 E-Line

An E-Line is one of the important and popular Carrier Ethernet service type that has been defined by the MEF. This technology is used for connecting exactly 2 UNIs, such that both UNIs can only communicate with one another. And therefore, an E-Line service is a Layer 2 communication technology that can provide dedicated P2P communication service. It provides P2P EVC pipe between two connecting UNIs at opposite ends and is used for Ethernet P2P connectivity between them.

E-line has been designed for organizations with single or multiple applications, not requiring a preset amount of bandwidth. And it is perfect for those organizations requiring physical separation of applications. At the same time, this technology can be used to construct services analogous to Frame Relay permanent virtual circuits. The EVC pipe can be set up with certain service attributes such as line-rate (granularity), delay, and so on (Figure 9.3).

Two types of E-Line services are as follows:

- Ethernet private line (EPL)
- Ethernet virtual private line (EVPL)

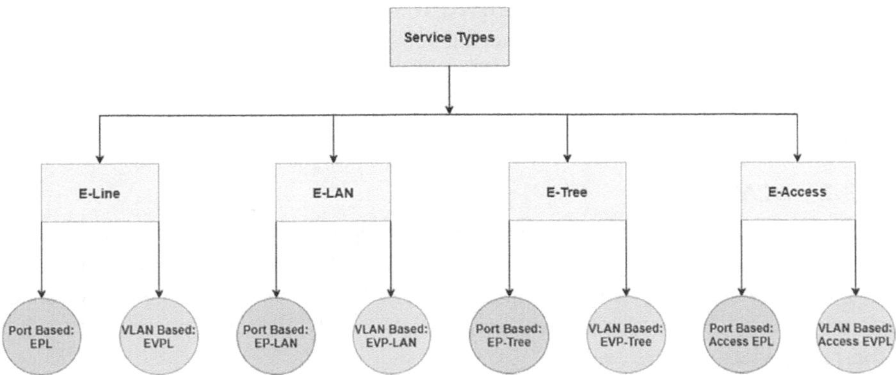

FIGURE 9.2 Different types of services of Metro Ethernet.

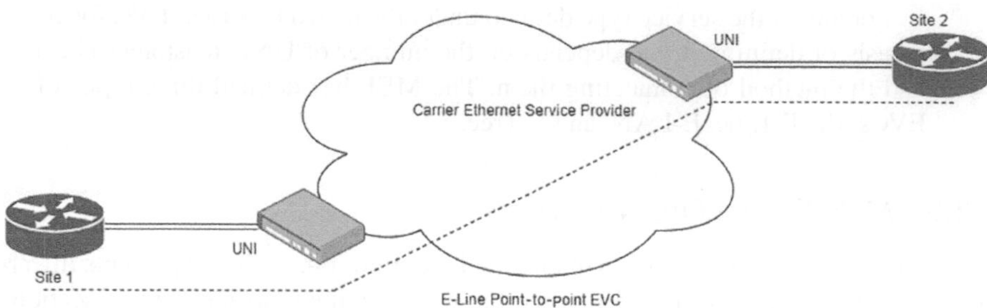

FIGURE 9.3 E-Line (point-to-point connection).

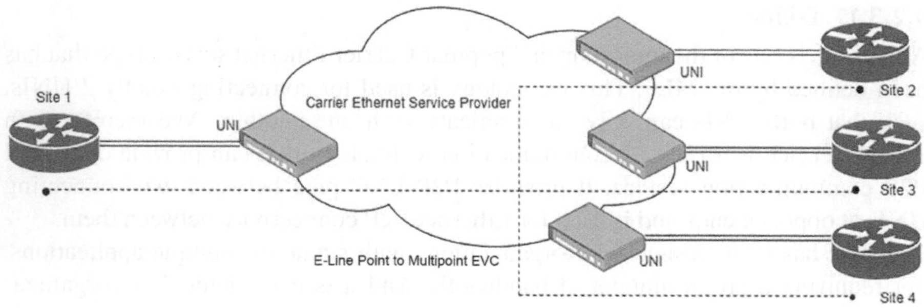

FIGURE 9.4 E-Line – EVPL service type (point-to-multipoint EVC).

Both services under the E-Line service type, the EPL and the EVPL, are data services defined by the MEF. Both services provide the simplicity of private connections combined with flexibility and scalability of MPLS across a range of bandwidths. However, the major difference exists between both service variants due to the configuration requirement to establish the private connection.

The EPL gives a P2P EVC between a pair of dedicated UNIs, providing a high degree of transparency. So, an EPL will have a dedicated P2P Ethernet connection, will require exactly two distinct and fixed UNIs for communication, and thus will provide high transparency for the Ethernet connection. On the other hand, an EVPL can enable multiple EVCs on single UNI, which can eventually support P2MP connectivity through the use of service multiplexing to provide one-to-many topology (Figure 9.4).

9.2.3.2 E-LAN

An Ethernet LAN or E-LAN service is a multipoint-to-multipoint data connection service that is used to connect a number of UNIs (2 or more) with providing a full-mesh connectivity to all those sites. This facilitates in establishing communication with each UNI to communicate with any other UNI that is connected to that Ethernet service.

This is suitable to those scenarios where we have a lot of sites and the requirement is for each site to be able to send frames directly to any other site with multiple

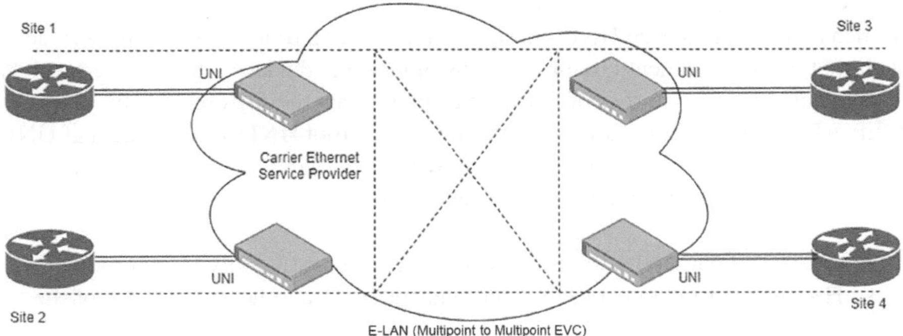

FIGURE 9.5 E-LAN service full-mesh topology.

locations requiring multipoint connectivity, and thus, we can use the E-LAN service that will work in a full-mesh topology and will act like a big switch. This offers a cost-effective solution, alternative to traditional network topologies that offer businesses with a more flexible and scalable network solution and provides connection of each location's premise equipment with enabling the transfer of data through a single Ethernet interface, with speeds up to 100 Gbps. The virtual private LAN service is another common name for E-LAN (Figure 9.5).

Two types of E-LAN services are as follows:-

- Ethernet private LAN (EP-LAN)
- Ethernet virtual private LAN (EVP-LAN)

As mentioned above, the EP-LAN and EVP-LAN are the two services categorized under E-LAN. An EP-LAN provides a multipoint-to-multipoint Ethernet WAN extending the LAN to a metro and is a port-based EVC connection service that allows seamless transmission of mission-critical applications and data across the entire network. The customer UNIs taking part in the EVC should have a fixed port with no service multiplexing allowed.

On the other hand, an EVP-LAN is categorized as an E-LAN service type, with an expectation of low frame delay, frame delay variation, and frame loss ratio. It is a multipoint-to-multipoint Ethernet service that provides a virtual connection and is basically a VLAN-based service which has been defined by the MEF, a Carrier Ethernet equivalent of virtual private LAN service or transparent LAN services, such that an EVP-LAN also has to facilitate with any-to-any communication between all customers locations associated with the customer's EVCs similar to the EP-LAN. EVPL and EVP-LAN service types are allowing service multiplexing that is allowed at the UNI and to share the same port with the customer VLAN IDs to be maintained across the network.

9.2.3.3 E-Tree

An Ethernet Tree or E-Tree service is another type of metro Ethernet services that is having the topology of a rooted multipoint service, which further connects a number of UNIs providing sites with hub and spoke multipoint connectivity. The topology is in such a fashion that each UNI is designated as either a root or a leaf. The service is such that each attachment circuit (each customer edge devices attached the service) is designated as either a root or a leaf and implements the feature as defined by the MEF, where the communication can happen from a root UNI to any of the leaf UNI, whereas a leaf UNI can communicate only with a root UNI. So, the data-forwarding rules followed by the E-Tree service are as follows:

- A leaf node or UNI will always send or receive traffic only from a root.
- However, a root node or UNI can send traffic to another root or any of the leaves.

The separation between UNIs is provided by E-Trees that is required to deliver a single service instance in which different customers (each having a leaf UNI) are connected to an Internet Service Provider (ISP) which has one or more root UNIs. It is useful for load sharing and resiliency schemes having more than one root UNI (Figure 9.6).

The following are the two types of E-Trees:

- Ethernet private tree
- Ethernet virtual private tree

The Ethernet private tree service is a port-based offering having fixed ports/UNIs for the root or leaf node having the connection to form a tree with rooted multipoint EVC. On the other hand, the Ethernet virtual private tree is a VLAN-based offering, like with Ethernet private tree service, each of the connected locations/sites or UNI is either a "root" or a "leaf". At the same time, since an Ethernet virtual private tree does not allow the UNI/ports to be fixed, multiplexing is applicable on per UNI basis.

9.2.3.4 E-Access

An Ethernet access service or E-Access service is an operator virtual connection (OVC)–based service which are defined in the MEF. An E-Access is supposed to

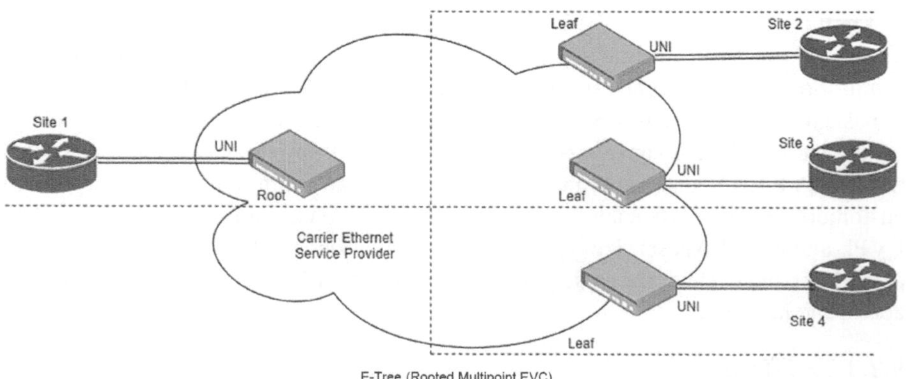

FIGURE 9.6 E-Tree service topology.

have at least one UNI OVC endpoint and one ENNI endpoint, which are OVC-based Ethernet services in contrast to the EVC-based services, where an OVC actually represents an association of OVC endpoints. Such that, when an EVC is spanning into multiple CENs, it is composed of segments in each CEN concatenated together to form an EVC, calling these segments OVCs. And thus, it will require an association of UNIs and ENNIs within a single CEN having at least one of these external interfaces. An OVC endpoint will be having the association of an OVC and an external interface.

The point-to-point E-Access services are defined – which are popularly access EPL, access EVPL, and access E-Line – containing exactly one UNI and exactly one ENNI.

9.3 CONCLUSION

Ethernet has been a very popular and most in-demand LAN technology that has covered a long journey after its innovation and has many specifications and developments since then. Ethernet technologies are commonly used in networking products that makes Ethernet products cheaper as compared with other networking technologies. It is due to certain benefits that Ethernet provides like low-cost and high-speed services because of which Ethernet has not been limited to LAN only. Metro Ethernet is one such technology that was the extension of traditional Ethernet (limited to a LAN) in metropolitan areas. In this chapter, we have explained about the concepts and some of the specifications of metro Ethernet technologies.

REFERENCES

1. Halabi, S., & Halabi, B. (2003). *Metro Ethernet*, Indianapolis, IN: Cisco Press.
2. McFarland, M., Salam, S., & Checker, R. (2005). Ethernet OAM: Key enabler for carrier class metro Ethernet services. *IEEE Communications Magazine*, 43(11), 152–157.
3. Huynh, M., Mohapatra, P., & Goose, S. (2009). Spanning tree elevation protocol: Enhancing metro Ethernet performance and QoS. *Computer Communications*, 32(4), 750–765.
4. Mirjalily, G., Sigari, F. A., & Saadat, R. (2009, December). Best multiple spanning tree in metro Ethernet networks. In *2009 Second International Conference on Computer and Electrical Engineering* (Vol. 2, pp. 117–121). Dubai, United Arab Emirates, IEEE.
5. MEF – Empowering enterprise digital transformation, Feb. 9, 2017. MEF, Accessed: June 21, 2021. [Online]. Available: https://www.mef.net/
6. Lam, C. F. (2003). *Modern Ethernet Technologies, Wiley Encyclopedia of Telecommunications*, Hoboken, NJ: John Wiley & Sons, Inc.
7. How Ethernet has enabled today's hyper-connected world, IEEE Transmitter, Feb. 25, 2020. Accessed: June 26, 2021. [Online]. Available: https://transmitter.ieee.org/ethernet-hyper-connected-world/
8. MEF Wiki Home Page – MEF Wiki, Mef.net. Accessed: June 28, 2021. [Online]. Available: https://wiki.mef.net/
9. Santitoro, R. (2003, April). Metro Ethernet services–a technical overview. *Metro Ethernet Forum*, Vol. 2006, 1–19, Available: https://www.mef.net/
10. Son, M. H., Joo, B. S., Kim, B. C., & Lee, J. Y. (2005). Physical topology discovery for metro Ethernet networks. *ETRI Journal*, 27(4), 355–366.

10 Modern Internet

ABBREVIATIONS

ARPANET Advanced Research Projects Agency Network
HetNet Heterogeneous networks
HTTP Hypertext transfer protocol
HTTPS Hypertext transfer protocol Secured
URI Uniform Resource Identifier
URL Uniform Resource Locator

10.1 INTRODUCTION

The Internet is seen as a widespread information infrastructure that has a global presence and has smoothened the process of information sharing from almost any city in the world to another. The history of Internet is complex and has its many aspects – technological, organizational, and community involved. The influence of Internet is not only limited to the technical fields of computer communications but has an impact throughout our society as we move toward increasing use of online tools to accomplish electronic commerce, information acquisition, and community operations.

The Internet that we know today is an ever-increasing worldwide network that has gone through several changes since the launch of its first workable prototype that came in the late 1960s with the creation of ARPANET, or the Advanced Research Projects Agency Network, which was originally funded by the U.S. Department of Defense. It used packet-switching technology that allowed multiple computers to communicate on a single network. The adoption of transmission control protocol/Internet protocol happened in the ARPANET in January 1983, and from there, the assembling of "network of networks" began by the researchers that became popular as the modern Internet.

The online world became more recognizable after the invention of World Wide Web by computer scientist Tim Berners-Lee in 1990. The success of Internet accessibility from user's perspective has a major contribution from the World Wide Web, and often the Web is confused with the Internet itself. The Web is a technology that has given meaning to the concept of "online" after the development of Internet, and therefore, it is actually considered just the most common means of accessing data online in the form of websites and hyperlinks. Considering all these facts, it is worth mentioning that the Web actually helped in popularization of the Internet among the general public, and served as a crucial step in developing the vast trove of data that we can access now on a daily basis.

The Internet of today has become very different compared with the original concept of the Internet at the time of development of its architecture and protocols around

DOI: 10.1201/9781003302902-10

the abstraction of communications between fixed end hosts. However, today's communication is characterized by the growing levels of mobility that devices are seeking. And thus, it is worth mentioning that the mobile wireless devices and the service end points in today's communication have outnumbered fixed end hosts with increasing levels of mobility. And therefore, it has a requirement of seamless support for mobility, and it is indeed a growing challenge to address this problem within the Internet community. Mobility has been addressed either within a limited environment (like a cellular network) or are inefficient when applied to the Internet (such as a MobileIP). Still, it is looking for scalable approaches to support mobility with the growing demand of mobile devices generating Internet data or the things looking for Internet connectivity.

Today's Internet is not only limited or restricted to the use of email or data communication but there are numerous Internet services on the top of Internet which are responsible of making it popular. These Internet services have impacted so much our lives that it has generated several markets and consumer base related to those services on the top of Internet. Example can be numerous and increasing streaming or over-the-top platforms, online music, online videos, online gaming, online conferences, online studies, online shopping, and several others. Both network infrastructure and services availability are essential as the scale and variety of network applications outweigh network speed upgrades. Such that, based on a survey only, the video content streaming represented 86% of global Internet traffic in 2016. But we already understood the fact that this Internet was originally and historically meant for transferring of data files, and thus had difficulties in providing highly available video services. At the same time, mobile Internet and the support for heterogeneous devices impose several challenges on achieving high availability of Internet. This had pushed the modern Internet to support extensible network infrastructure, support the connectivity of versatile network devices, and on the top of that numerous uninterrupted Web and mobile services.

As with increasing demand of services and numerous services availability on or based on Internet, information security is a growing challenge as well. Businesses running on the top of Internet today have millions of user data, personally identifiable information and others which should be protected from leak and need to be kept and maintained with data integrity. The Internet infrastructure must be supportive to contain and eliminate promptly the security threats to minimize their adverse effects. The modern Internet architecture is required to be resilient and to be able to recover from faults so as to maintain its high availability. Such that, loss of Internet connectivity is not an affordable matter to the businesses running on Internet as a connectivity loss of minutes may result into loss of a million in some cases.

We have already learned about the story of Internet in our earlier chapter and also have gone through in detail description of the ARPANET. In this chapter, we will see the latest and recent technological innovations and support that the modern day Internet can provide.

10.2 PRINCIPLES OF MODERN INTERNET ARCHITECTURE

Every architecture is based on certain design principles that conceptually works as a basis in the formation of that architecture. From the decades-long evolution journey of the Internet, its architecture is something which is considered as progressively moving from the functionality of pure network connectivity to a networking ecosystem. The modern Internet architecture integrates the network connectivity with the several other features, services and functionalities that combines networks, computing, storage, security, and automation. And thus, it is important to understand the design principles governing modern Internet that was limited earlier till the functionality of network connectivity [1]. Here, we are going to discuss some of those design principles of modern Internet architecture that is not only driving most of the engineering decisions at conception level but also working at operational level. Design principles of Internet architecture become very important and obvious while designing and formulation of the Internet because of the fact that it provides fundamental characteristics of the process to guide the designing of its protocols. At the same time, we should also remember another very important fact that that the technical change is a progressive and continuous journey in the overall information and communication technology industry that also includes architectural principles of the Internet [2].

10.2.1 Heterogeneity

Heterogeneity is one of the important parameters of modern Internet architecture. Such that, a heterogeneous network can have interconnection of different types of nodes with different types of links. Heterogeneous network in computer networking is such a network that connects computers and other devices where the operating systems and protocols have significant differences with each other. These interconnected nodes of the network may have rich information, and it can enhance nodes and links mutually, at the same time the information can propagate from one type to another. Heterogeneous networks or "HetNet" is also very popular in wireless networking that consists of different access technologies and is called a wireless heterogeneous network. A HetNet generally refers to the use of multiple types of access nodes in a wireless network and can have a combination of various cellular accesses such as macrocells, picocells, and femtocells to provide wireless coverage in an environment with a wide variety of wireless coverage zones.

Heterogeneity is the need of modern networking, and it is inevitable and must be supported by the network architecture. Heterogeneity can be considered at many levels in the modern Internet architectures, which are listed below:

1. In terms of supporting numerous application-layer services and the traffic mix in the network overlay
2. In terms of network management and control
3. In terms of the transport layer mechanism and protocols
4. At Devices and nodes level
5. In terms of Scheduling algorithms and queue management mechanisms
6. In terms of Routing protocols
7. In terms of multiplexing
8. In terms of protocol versions and implementations
9. At the underlying data link layers

10. At the physical layers
11. In terms of congestion at different times and places

Heterogeneity is expected to be increasing day by day in the network, and with the evolution and support for the Internet of Things, it will get more and more diverse and heterogeneous in future. Such that, modern Internet coexists with multiple types of hosts/terminals, network nodes, communication methods or technologies, protocols, and applications.

10.2.2 Scalability

Scalability is one such feature that was adapted since the beginning of the evolution of Internet. Scalability defines that particular capability of a system, network, or process with that it can handle the growing amount of work, or its potential can be increased up to the required level in order to accommodate that growth. Scalability refers that the designs should be supportive to readily available with as many nodes per site and with numerous sites as per the increasing demand of the network. This principle is therefore important for the Internet because an Internet is the global giant network (network of networks) and its architecture should scale invariantly to fulfill the global design requirement. Modern Internet has to support increasing number of devices with Internet access (computers, mobile devices, etc.), communication nodes, autonomous systems, and applications and their number is expected to increase significantly in future. With the growing demand of the support for Internet of Things, the requirement of direct interconnections of sensor networks with the Internet is increasing exponentially and pushing for more number of Internet nodes. Therefore, scalability is one such the principles of design architecture of modern Internet that govern its usability.

10.2.3 Simplicity

Simplicity is another principle of modern Internet architecture that is very desirable for Internet's flawless designing and smooth operation. Being versatile and diverse at one hand, the architecture should follow a simplistic rule to provide flexibility. It is a general understanding that complex systems are usually less flexible and reliable. Complexity in terms of architecture refers that, it becomes mandatory minimizing the number of components in a service delivery path for increasing the reliability that may include a protocol, a software, or a physical path. However, the principle of simplicity has been challenged sometimes as well as of the fact that, a complex problem has more elaborated solution sometimes and a multidimensional problems like the Internet architecture can provide a nontrivial functionality. Considering one general complexity problem of the Internet architecture as follows: determining the placement and distribution of functionality to globally minimize the architectural complexity. And thus, lowering of complexity over space arbitrarily considering this agenda can result into local minimum which can be detrimental globally. Hence, the scalability and simplicity are such principles, which should be handled as strongly interconnected first priority design principles when designing the Internet.

10.2.4 Robustness and Adaptability

The principle of robustness is also known as Postel's Law in computing after Jon Postel, whose wording was used in the early specification of transmission control protocol. According to this law, each of the protocol implementations must interoperate with others which are created by individuals differently. It is one of the design guidelines that states that each one should "be conservative in what you do, be liberal in what you accept from others". This has been reworded as "be conservative in what you send, be liberal in what you accept" as the same protocol may have different interpretations. The purpose of this principle is to maximize the interoperability between different implementations of similar protocol, specifically in those circumstances where the underlining protocol may have ambiguous or incomplete specifications.

At the same time, the concept of adaptability represents that the network should be adaptable against different packets designed to worsen the network. Therefore, the Internet designers should consider it in the designing phase itself that the network can be filled with malevolent entities to send packets for the worst possible effect in the networks. Consideration of this assumption gives our modern Internet suitable protective design, and as a result, protocols would improve their robustness. The modern Internet is expected to adaptable and robust to cope up with different types of network managed by different organizations generating suitably varying applications data with minimization of malfunction, uninterrupted operation and interoperability. Modern Internet should be capable to handle mission and time critical applications, related to health, energy, transport, digital marketing, stock markets, supply chain etc. As a result, the principle of network robustness and adaptability becomes one important aspect of modern Internet architecture.

10.2.5 Loose Coupling

Generally, in the software and computing systems, coupling is considered as one important parameter for interconnecting different components, systems, or networks, which determines the dependability between each architectural module and component that depends with another module or component. Loose coupling is such a method that defines the dependability of least extent practicability between interconnecting system components on each other. A loosely coupled system is such that its components or modules make use of little or no knowledge of other individual components. Between the interconnecting systems by considering the maximum number of change in the components occurring without adverse effects the extent of coupling can be measured. In the modern Internet architecture, modularity with loose coupling is best preferable and the decoupling between applicative layers is a good example representing loose coupling in the communication stack. The degree of coupling varies, and they often show nonlinearity in terms of interaction as with things get larger, generally there is increased interdependence between components. The preferable mode is of loose coupling because it minimizes the unwanted interaction among system elements. However, sometimes it can also create difficulty in establishing the synchronization between different components when that interaction is desired.

Loose coupling represents an important consideration for a well-structured and well-designed system architecture as of the fact that is mentioned below:

1. It has simplified testing and troubleshooting requirements because the problems can isolate easily and remain locally at the same time it is unlikely that one faulty system will make faulty other systems due to less system interaction.
2. It helps to achieve high readability and maintainability.
3. It helps in minimizing the unwanted interaction between several components of the system.

However, when we compare a tightly coupled system with respect to the loosely coupled system, it is very likely to experience unforeseen failure states comparatively. Generally, complex interactions between the several components of a tightly coupled system increases complexity in the system design and make the system hard to understand and predict, which in turn provides less flexibility to the system in recovering from failure states. And thus, in the modern Internet architecture as well, it became critical to include and preserve the principle of loosely coupled system design as a result of the increasing importance of the availability objective.

10.2.6 Naming and Addressing

Address space in computing is considered as a range of discrete addresses where each of the addresses from that address space may represent a network host, peripheral device, disk sector, a memory cell, and any other physical or logical entity. The number of the addresses available in an address space usually depends on the underlying address structure and the address spaces are created by combining such uniquely identified qualifiers that makes an address unambiguous within the address space. One of the most common features of the address spaces is address mapping and translation, which means that some higher-level address has to be translated into lower-level ones for some reason. Examples are the Domain Naming System that maps the names to network address, that is, IP, Address Resolution Protocol, and Network Address Translation. Naming and Addressing is ever existing feature of Internet or any network as well, and the modern Internet architecture has to have this functionality. This functionality provides the identification of the connected devices on the network as well as naming convention for resolution of the applications hosted from names to actual server address (IP). It is a common functionality of the Internet where the upper layer Internet protocols should unambiguously identify the application end points and should be hardware medium independent, whereas the hardware addressing should be capable of facilitating any new digital transmission technology along with decoupling the hardware from its addressing mechanisms. It is such a technique that helps the Internet to easily interconnect with different transmission media, offering a single platform to support a wide range of applications on Internet, its diverse infrastructure, and services.

Considering the increasing demand of devices needing for Internet connectivity, the highly scalable core Internet, the modern era of Internet needs to cope with the use of name and address spaces following augmentations are considerable:

1. Need to avoid such designs requiring hard coded addresses or to be stored on non-volatile storage, also a discovery process is recommended.
2. Common naming structure should be used in the design.
3. Design should ensure that the Locators (LOC) and Identifiers (ID) are separate.

It is supposed that in future the end points (ID) and their respective attachment points (LOC) need unambiguous naming and addressing, which should be unique within the scope.

10.2.7 Distributed Architecture

Networking means connectivity between nodes which are distributed geographically. Since the beginning, the architecture of Internet has been of a distributed as with its classic use case at the time of implementation of ARPANET was of sharing computational resource among several research institutions. Its design and architecture have been major technical achievement in past that enabled program running anywhere to address messages to program anywhere else. Because the Internet consists of globally linked enormous number of big or small computer networks and thus the architecture of Internet will always considered as distributed. It consists of an enormous number of interconnected Intranets belonging to numerous organizations which wants their services, applications, or resource to be distributed globally over the Internet. These Intranets are connected by very high-speed backbone network which is called as the core of the network. Modern Internet is actually such a giant network worldwide with vast interconnected collection of computer networks of many different types and with range of types always increasing.

10.3 THE WEB

World Wide Web also popularly mentioned as the term "Web" represents the pages that we see on our devices when we are online [3]. Web is something that has popularized another term in relation to Internet and digitalization that is "online" or "things being online". Internet is that entity which provides the infrastructure of network that supports Web works, as well as what emails and files travel across. We can think of the Internet as those roads that give connectivity to towns and cities together and handle Web traffic from one location on the globe to another. In the success story of Internet, a major role and credit has to give to the Web. In the modern era of Internet the Web has also become more interactive, dynamic, analytics driven contents and agile. Modern Web is also getting clubbed with and working in coordination with Artificial Intelligence to enhance user experience.

Tim Berners-Lee, an English scientist, invented the World Wide Web in 1989 and wrote the first Web browser in 1990. After that, the Web began to enter into everyday use case with the availability of the Internet. Web is considered since then as the core to the development of Information Age with the evolution of Internet [4].

The success of World Wide Web is considered because of the design of its software architecture to meet the needs of an Internet-scale distributed hypermedia application. Scalability of component interactions, interface generality, independent deployment of components, reduction of interaction latency, security enforcement, and encapsulation of legacy systems are some of the features on which modern Web emphasizes.

Some of the terminologies related to Web are as follows:

URL: A URL or Uniform Resource Locator is termed as a Web address that represents a Web resource location on the computer network (Internet) and mechanisms for retrieving those resources.

URI: A URI or Uniform Resource Identifier is something which is used to identify a resource and differentiate it from other resources by the use of the name of the resource or the location of the resource.

Hypertext: Hypertext is those text or word on the webpage that are used to contain a link to a website.

Hyperlink: A hyperlink is a reference to the data that the user follow by clicking or tapping on it, it is a word, phrase or image that can be clicked on to jump to a new document.

10.3.1 Web vs. Internet

Web also represented as World Wide Web is often confused with Internet most of the time, and both are used without much distinction. However, both of these terms do not have the same meaning. Internet as we discussed earlier is a global giant network of interconnected numerous computer networks. On the other hand, the World Wide Web can be considered as a global information system using hypertexts that include documents and other many resources, which are linked with hyperlinks and URI. These resources on the Web are accessed using HTTP (Hypertext transfer protocol) or HTTPS (Hypertext transfer protocol secured) protocols from the web browser on your computer or end device. These are application-layer protocols used for Internet's transport protocols.

We view a web page existing on the Web either by typing the URL of that page into the web browser of our computer system or by mouse clicks on the hyperlink to that page or resource. Following that, the web browser has to initiate a sequence of background communication and messages to fetch and display the requested page. This process of browser based information processing through Web or accessing the view web pages or following the web resources through hyperlinks is known as "browsing" "web surfing" (after channel surfing), or "navigating the Web".

10.3.2 Web 2.0

Web 2.0 describes the current state of the Web, which refers to websites emphasizing user-generated contents, ease of use, participatory culture, and interoperability. It has more contents generated by the users and having more end-users usability compared with the earlier Web. Web 2.0 refers to those Web-based Internet applications and usability that have transformed the digital era. Darcy DiNucci coined the term "Web

2.0" in 1999, and it was further popularized by Tim O'Reilly and Dale Dougherty at the first O'Reilly Media Web 2.0 Conference in 2004 [7].

It is worth mentioning that Web 2.0 [5–8] does not have a hard boundary with Web 1.0 (also Web 2.0 and Web 1.0 are not represented like other software systems as different software versions), but rather, have a difference in the understanding, practices, and principle. As we discussed that, these are not the software versions although the term Web 2.0 mimics the numbering of software versions. And thus, it does not represents as such a formal change in the nature of the World Wide Web. Web 2.0 can be visualized as a set of principles and practices that tie together to merely describe a general change that occurred during this period as interactive websites proliferated and came to overshadow the older, more static websites of the original Web.

Websites following the principles and practices of Web 2.0 facilitates user interactions and collaborations with each other (through social media dialogue) as the creators of user-generated content in the community (Web domain). This contradicts with the past Webs (Web 1.0-era websites) where the contents were limited to user in a passive manner. There are numerous examples of Web 2.0 around us which can be considered as one of the digital disruptions on top of modern Internet.

Some Web 2.0 features are mentioned below (but are not limited to below only):

- Social networking sites or social media sites (e.g., Facebook)
- Blogs
- Wikis
- Folksonomies ("tagging" keywords on websites and links)
- Video sharing sites (e.g., YouTube)
- Image sharing sites
- Hosted services
- Web applications ("apps")
- Collaborative consumption platforms

There have been several criticism of the term as well, critics claim that "Web 2.0" does not represent a new version of the World Wide Web at all but merely continues to use technologies and concepts of so-called "Web 1.0". Mentioning that ideas of Web 2.0 were already featured in implementations on networked systems well prior to the emergence of the term "Web 2.0". Web 2.0 being substantially different from prior Web has been challenged by Tim Berners-Lee the inventor of World Wide Web. He has been an outspoken critic of the term Web 2.0, describes the term Web 2.0 as jargon to World Wide Web, and mentions that the original vision of the Web was a collaborative medium.

10.4 CONCLUSION

Modern Internet is not limited to any specific region in a country but a global network connecting millions of computers worldwide. It is a decentralized network where each Internet computer is independent and can communicate with any other computer as long as they are connected to the Internet. Heterogeneity is an important feature of modern Internet architecture and Internet is considered as the largest example of a distributed system. A heterogeneous network is the interconnection of different types of network that connects computers and other devices with having different types of operating systems and protocols with each other. It is worth mentioning that the nature of heterogeneity is one of the main factors for popularity of Internet. The Internet itself is actually a giant network that is composed of millions of small subnetworks with the possibility of each of the subnetworks using different protocol for communication and data transfer within the network.

World Wide Web is an information-sharing model built on top of the Internet to access information over the medium of Internet. It uses the HTTP protocol for communications between the applications to exchange business logic and to share information. The combination of the Web and the Internet is one of the most important factors behind the digitalization of technologies in modern era. Web is providing the platform to host and access several services and Internet is providing the medium for communication. Web services hosted on any part of the world with the help of innovative technologies via the Internet can be quickly accessed in another or very distant part of the world.

REFERENCES

1. Papadimitriou, D., Zahariadis, T., Martinez-Julia, P., Papafili, I., Morreale, V., Torelli, F., … & Demeester, P. (2012, May). Design principles for the future internet architecture. *The Future Internet Assembly,* 55–67. Berlin, Heidelberg: Springer.
2. Fielding, R. T., & Taylor, R. N. (2002). Principled design of the modern web architecture. *ACM Transactions on Internet Technology (TOIT),* 2(2), 115–150.
3. Wikipediacontributors, World Wide Web, Wikipedia, Accessed: July 5, 2021. [Online]. Available: https://en.wikipedia.org/w/index.php?title=World_Wide_Web&oldid=1073557476
4. Berners-Lee, T., Cailliau, R., Luotonen, A., Nielsen, H. F., & Secret, A. (1994). The world-wide web. *Communications of the ACM*, 37(8), 76–82. https://doi.org/10.1145/179606.179671
5. Lewis, D. (2006). What is web 2.0?. *XRDS: Crossroads, The ACM Magazine for Students*, 13(1), 3–3.
6. Yu, C., & Du, H. (2007). Welcome to the World of Web 2.0. *The CPA Journal*, 77(5), 6.
7. O'Reilly, T., & Dougherty, D. (2004). O'Reilly Media Web 2.0 Conference. Recuperado el, 7.
8. O'Reilly, T. (2007). What is Web 2.0: Design patterns and business models for the next generation of software. *Communications & Strategies*, 1, 17.

11 Software-Defined Networking

ABBREVIATION

API	Application Program Interface
NBI	Northbound interface
NOS	Network Operating System
OF	OpenFlow
SBI	Southbound Interfaces
SDN	Software-Defined Networking
TCP	Transmission Control Protocol

11.1 INTRODUCTION

While discussing about the modern networking, it will take us through the concepts and technologies of networking, which became very popular in the modern era that is the software-defined networking or SDN. As it is called the software-defined networking (SDN), it is such a networking technology where the network services can be created or defined based on the software configuration [1–5]. SDN is a dynamic technology that is manageable, cost-effective, and adaptable, and that makes it ideal for modern-era applications which have a dynamic nature of work and high-bandwidth requirement. The scope of SDN remains in making the network architecture with a software-centric approach to decouple the network control (control plane) and forwarding functions (data – forwarding plane). This happens in such a manner that the network control becomes dynamic, programmable, and centralized policy control, whereas the underlying infrastructure is abstracted for applications and network services with no manual dependence on physical infrastructure. In general, the network architecture has traditionally the networking devices with combined control and data plane functions into a single hardware entity, typically a router or switch. The control plane is such a network element that determines the interaction among devices within a network. The routing protocols like Open Shortest Path First (OSPF) and Border Gateway Protocol and switching protocols like Spanning Tree Protocol are the control plane protocols that are used, respectively, at routing and switching levels. Using these device-centered control plane protocols, a device determines the suitable device port or interface that is used to forward data at the data plane. In that way, the use of SDN has shifted the control plane from being device centric, that is, at the individual device level to overall network centric.

With SDN, the importance of software-centric approach to network services has increased, which was traditionally or initially mostly having the hardware-centric approach. This has been more beneficial with giving several benefits to the enterprises,

DOI: 10.1201/9781003302902-11

network service providers, device-vendors, organizations, etc., such as cost-effective, dynamic, software-controlled, and less manual effort. The use of software in controlling the network services has developed a virtualization layer on top of the hardware infrastructure. This virtualization has helped in developing the network services based on the requirement of applications and not directly depending on the underlying hardware infrastructure. Thus, in simplest manner, we can say that an SDN is an evolution of the networking technologies and is the modern-era networking with a new approach to networking as compared with the legacy networking, which has given a framework for modern application-centric networking with multiple solutions available.

11.2 ARCHITECTURE OF SDN

The architecture of SDN at the high level is seen as the abstraction of three separate planes representing three different layers with their respective functionalities, and the networking through SDN is the outcome of the interaction among these layers. These layers of the SDN architecture are the application layer, the control layer, and the infrastructure layer. While designing the architecture of SDN, the goal was to provide open interfaces to enable the development of software that will help in controlling the connectivity provided by network resources and the flow of network traffic though them with a possibility of traffic inspection and modification that may be performed in the network. Please be noted that the layers mentioned below or anywhere in the SDN should not be confused with the layered network architecture we have seen in traditional networking or should not be considered in relation to the open system interconnection models or TCP/IP layers.

11.2.1 Components of SDN

Considering the SDN architecture mentioned in Figure 11.1, the following are the components of the SDN that we are going to discuss now.

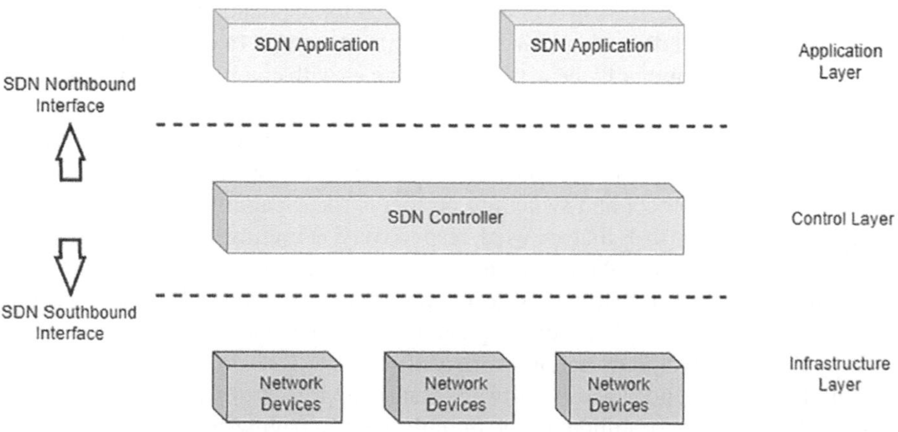

FIGURE 11.1 SDN architecture.

11.2.1.1 Application Layer

Network application layer or the application plane is at the top of the architecture, which consists of the numerous SDN applications possessing some network services and will be leveraging specific APIs to call the respective network service. Application layer has to communicate downstream with the control layer or control plane through their respective interfaces and is further translated to the Infrastructure layer to accomplish the data-forwarding functions.

11.2.1.2 Northbound Interface

The northbound interface (NBI) in SDN is an abstraction that allows the communication between the control planes with higher layer, that is, the network applications. Diagrammatically, the northbound traffic flow can be thought of as going upwards, and thus, they are mentioned in the upper part of the SDN architecture so as to connect with the top layer, conversely to the southbound interface that remains in the bottom part of the architecture. Mostly, the northbound interfaces are made using a software system, where applications are built programmatically or using APIs such as RESTful API. The supporting programming languages can be Python, Java, and many more to enable faster development, lower investment costs and easier troubleshooting. The communication between the applications and controller using northbound interfaces will be bidirectional such as the controller will notify to the application at the top layer of the proceedings happening in the network underneath. It may also consists of several types of events constituted under separate packets established by the controller or state alteration in the topology, such as a connection going down. At the same time, the applications can have different approaches as well in response to the events received that may comprise of dropping, modifying or forwarding of the packet for a received packet event.

11.2.1.3 Control Layer

The middle layer in the SDN architecture is the control layer or the control plane that consists of an SDN controller device. The control layer has its upper boundary with the application layer through the interconnections using SDN northbound interface. The lower boundary is with the infrastructure layer through the interconnections using the SDN southbound interface. This layer provides the control over the network elements through a single pane of glass. SDN controller has to translate the networking requirements based on the SDN applications to the bottom infrastructure layer via the southbound interface. Controller has to perform the network orchestration requirement as per the SDN application demands with pushing the policy controls to the lower-level network elements. The controller provides a programmable interface to the network admins by using that a network admin can configure or program the policy control specific to the need of SDN application.

The SDN control plane can be implemented in multiple ways, either in a centralized, hierarchical, or decentralized design. Initially, the proposals for the implementation of SDN control plane were mostly focused on the centralized solution. In a centralized solution, a single control entity has a global network view. The centralized control solution simplifies the implementation of the control logic in

SDN, but at the cost of scalability issue as with increasing the size and dynamics of the network. And thus, several of the approaches have been proposed in the literature in order to overcome the limitations possessed by the centralized control that generally belongs to two categories, hierarchical and fully distributed approaches. In hierarchical solutions, the entire control plane is divided into certain network segments with each network segment working under controllers distributed over those network partitions. These distributed controllers are further placed into hierarchical order so that a logically root controller can take those decisions that require network-wide knowledge. In distributed solutions, the entire control plane is placed with controllers distributed overall and these controllers operate on their respective local view. These controllers also exchange synchronization messages to enhance their network information.

Placement of controller is very important design problem while designing a distributed SDN control plane, and the number and placement of control entities will depend on that design. Some important parameters that are considered while designing of the control plane are propagation delay between the controllers and between the network elements, control path reliability, fault tolerance, and application requirements.

11.2.1.4 Southbound Interface

Unlike the northbound interface which connects the higher layers in the architecture, the southbound interfaces (SBI) are those interfaces that are used to connect the lower layers in the SDN. The southbound interfaces are the linking connections between control plane and network entities (infrastructure layer) and therefore a southbound interface has to explain the entire communication procedure overlapping between the network devices underneath and the control plane. This communication will depend on the set of guiding rules or the protocol established for the control plane's interaction with the substructure physical or virtual infrastructure. At the same time, it is considered as an industry standard to justify the ideal approach for the SDN controller to communicate with the forwarding plane. That will further help in modifying the network as per the application and would let it progressively move along with the advancing enterprise needs. Network admins can add or remove entries present in the internal flow table of network elements (like switches and routers) for composing a more responsive network at the Infra-layer for real-time traffic demands.

11.2.1.5 Network Operating Systems

"NOS" or network operating system is an important concept and core element in SDN architecture, and it is used for the abstraction of infrastructure and feature-rich protocols. It further helps in getting a general control plane and a unified protocol operating view. The provision of NOS in conjunction to SDN has to orchestrate the design requirement of shifting control from the specific network functions and vendor-independent implementation of control plane. It will also include the extension of abstraction from computing process locally inside a network device for data-forwarding operation. With NOS abstraction at the controller level will abstract the details of the SDN controller-to-device protocol that will facilitate in the applications to communicate with those SDN devices without knowing the difference.

11.2.1.6 Infrastructure Layer

The bottom layer in the SDN architecture is the Infrastructure that consists mostly of the Network devices. It is that layer where the actual data-forwarding actions takes place based on whatever data-forwarding policy is pushed by the controller. This layer is also sometimes called as the data plane in SDN architecture due to the same reason as of the data-forwarding and -processing action happening at this layer.

11.2.2 Traffic Flow in SDN

Considering the analogy from traditional data center networking, there are two popular traffic flows in SDN as well which are mentioned below, respectively, the north-south flow and the east-west flow. Based on the plot of the data center network and the direction of the traffic flow, the segmentation traffic was done for the purpose of a better control, management, and security of the organization's network.

11.2.2.1 North-South Flow

In the data center networking, the north-south traffic is usually represented to describe the client-to-server traffic moving between the servers located at the data center and the client sites located outside of the data center. Diagrammatically, the north-south traffic is also depicted as a vertical traffic flow to illustrate that the flow of traffic is above or below the data center. Such that, the traffic that enters the data center through its perimeter devices is said to be the southbound and the traffic exiting via the perimeter devices is said to be northbound. The north-south flow is also called as the mice flow in SDN as they are short size payloads, but they represent about 90% of the entire flows.

11.2.2.2 East-West Flow

As per the traditional networking, the east-west traffic is quite popular and often used to represent the transfer of data packets from one server to another server within a data center. The idea behind the term east-west flow for such type of traffic comes from pictorial representation of the network diagram that usually shows LAN traffic horizontally. As with the segmentation of the network architecture for ease in network control and management, it is very obvious to keep all those segments in sync to each other. This has generated the need of east-west flow in the network. The volume of east-west traffic has grown comparatively from past few years that has been seen as a result of virtualization and converged infrastructure inside the data center. With the evolution of the data center technology today, various functions and services are running virtually that previously ran on physical hardware such as virtual firewalls, load balancers, network controllers, virtual machines, and other software-defined networking (SDN) concepts. As such, the network has to experience very high and increasing volume of traffic due to the relay of data from

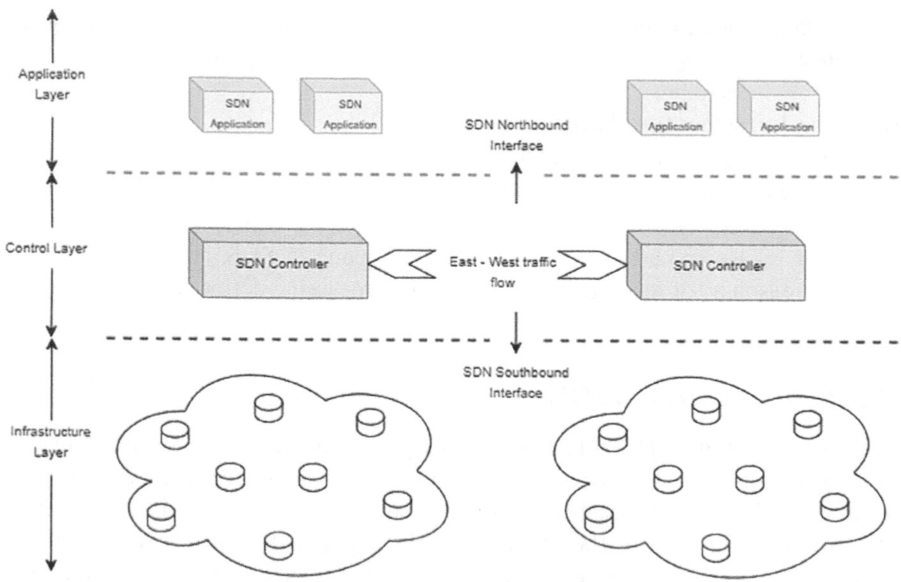

FIGURE 11.2 East-west flow in SDN.

these components to each other, which can also cause the increase in latency and congestion in the network. The east-west traffic is critical for the control and management point of view, and at the same time, it is very important for the security inside the data center network for the insider threat, malwares, intrusion happened inside the network.

The concept of east-west flow is also very popular in the SDN as well, and in simple manner, it can be understood as the traffic flow happening at the controller level from one controller to another as shown in Figure 11.2. The east-west flows are also mentioned as elephant flow in SDN due to their bulky size in the volume, which accounts for around 80% of the payload volume. However the number of elephant flows in the data center network remains around 10% on average of the total traffic flows.

Considering the need for east-west flow in a network and to compensate the demerits in traditional networks, many organizations have migrated their network architecture from traditional three-tier data center approach to various forms of leaf-spine architectures or two-tier approach. The three-tier architecture as mentioned in Figure 11.3 consists of three separate layers of devices arranged, respectively: the core layer, the aggregation layer, and the access layer. The leaf-spine approach as mentioned in Figure 11.4 is observed to be well-suited to handle higher volumes of east-west traffic than compared with the three-layer approach. Such that, the leaf switches are front-end switches used to consolidate traffic from users and then connect to the spine. The spine switches are in the core of the network which comprises core servers and storage systems.

SDN gives another aspect of control and management to east-west traffic and organizations deploying the SDN on a leaf-spine fabric can take advantage in maintaining the overhead for the east-west flows and making east-west traffic management more efficient.

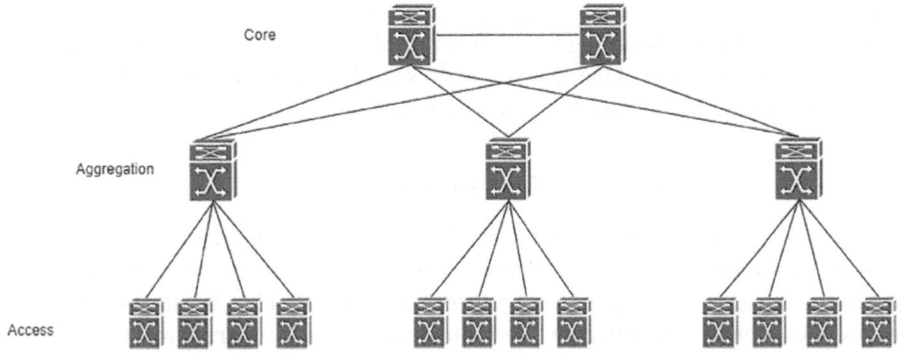

FIGURE 11.3 Traditional three-tier data center architecture.

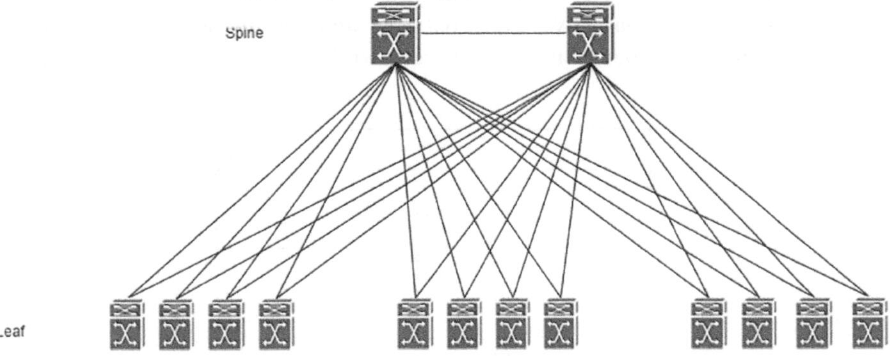

FIGURE 11.4 Data center evolution with leaf-spine architecture.

11.3 OPEN FLOW PROTOCOL

OpenFlow (OF) [6, 7] is the most popular and commonly used protocol standard for the implementation of SDN. It is used in managing the southbound interfaces of the SDN architecture and the first standard interface defined to facilitate interaction between the control and data planes. The SDN controllers are required to communicate with the network routers and switches via the Southbound APIs, where OpenFlow protocol is used to provide software-based access using the flow tables to instruct the network routers and switches to direct network traffic. The devices such as routers and switches of the network can be both physical and virtual (hypervisor based). A number of the device manufacturers and vendors worldwide are supporting the SDN by manufacturing these routers and switches, which can be controlled by SDN controller using the OpenFlow standard interface. Through that software-based control, the required policies are pushed into the underlay network such that using these flow tables, an administrator can quickly change network layout and flow of traffic. The OpenFlow protocol also provides a basic set of management tools that may be used to control features such as changing of topology and packet filtering. The Open Networking Foundation is the nonprofit organization that defines, controls, and manages the OpenFlow protocol standard and is working for the adoption and promotion of the software-defined networking. The Open Networking Foundation is led by the board of directors from seven companies owning some of the largest networks worldwide such as Deutsche Telekom, Facebook, Google, Microsoft, Verizon, Yahoo, and NTT.

11.3.1 OpenFlow Table and Flow Entries

From the higher application layer, the SDN controller takes the information using the NBI and converts them into flow entries, these flow are further fed to the switch at the lower/Infra layer by OpenFlow protocol through SBI. The OF standard can also be used for monitoring switch and port statistics in network management.

As mentioned in Figure 11.5, the OF protocol is only established between SDN controller and the OF switch, and the rest of the network remains unaffected. The OF protocol is working on top of the TCP where TCP 6633 for the earlier versions OF v1.0 and TCP 6653 for OF v1.3 later onwards. To establish the OpenFlow channel based on TCP connection between the controller and the OF switch, it will require first an IP connectivity between the controller and the switches. The establishing of

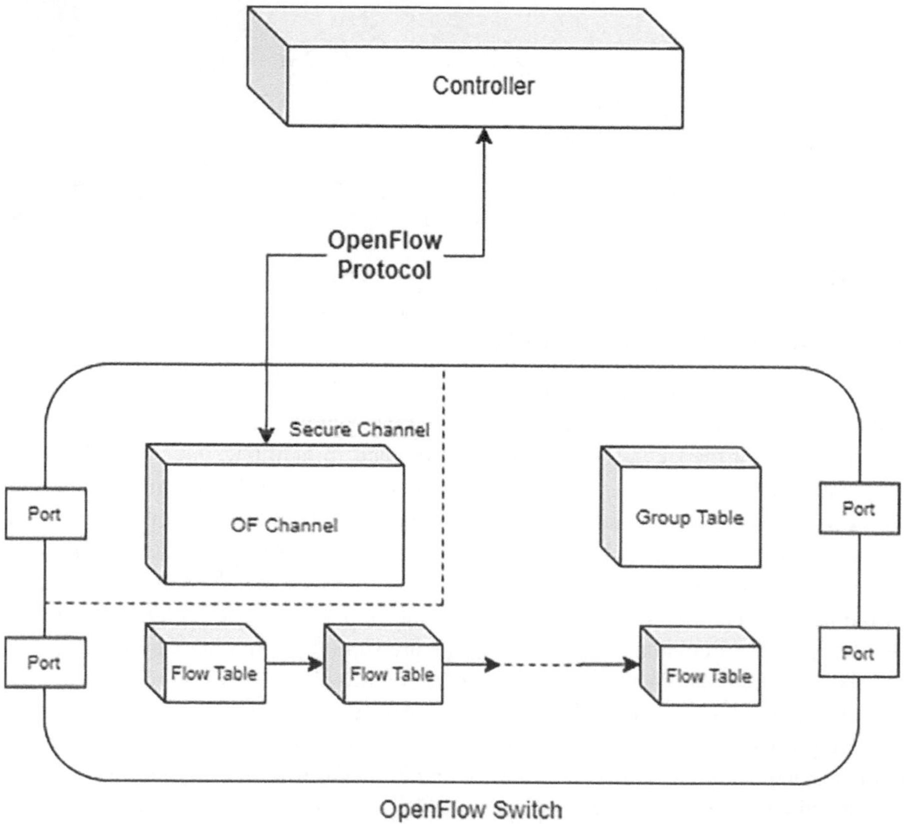

FIGURE 11.5 OF protocol in the SDN.

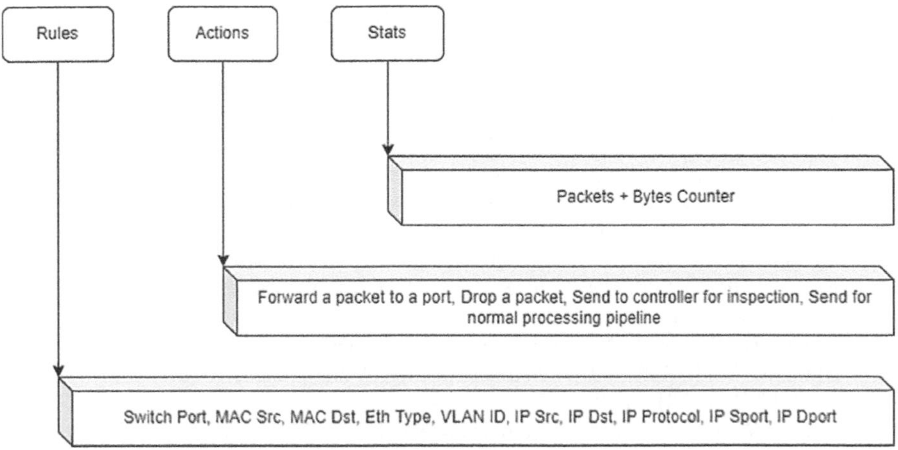

FIGURE 11.6 Diagram representing the flow table of Open Flow.

the OF channel is based on a successful three-way handshaking, this connection is very essential as it is the only way for a switch to communicate with a controller and forms a secure channel (Figure 11.6).

Flow table store the details of the flow entries useful for the packet forwarding that tells the SDN switch about an action to take when a packet will arrive at the switch port. The flow entries have details such as IP address, port number, MAC address, and VLAN IDs. Packets forwarding happens in the network as per rules in the respective switch flow tables. When a packet will arrive at the switch port then the switch will match specific parameters and select the best matching flow entry from the flow table to execute the action associated with that flow. The actions can be forwarding the packet to some other port, dropping a packet, flood a packet or send the packet to the controller for further inspection. Further to this, multiple components are present as well in the flow tables like match fields, priority, counters, instructions, timeouts, cookies and flags in a flow entry during a flow matching process.

11.4 CONCLUSION

SDN has emerged as the industry's response to meeting the challenges of the demands from a network that increases network's automation, allowing networks to react dynamically to changes in usage patterns and availability of network resources. It helped in adjusting network architectures instantly, managed and controlled through a software-based API, respond to application and user requests, and services can be introduced far more quickly, easily and at a lower cost.

The use of SDN into the enterprise networks is a better way to manage the WAN for connectivity to private Cloud deployments by optimizing utilization and controlling costs. The adaptation of the concepts of SDN within the enterprise network to achieve WAN connectivity is very popular and is referred to as SD-WAN or software-defined wide area network. Software-defined wide area network has all the features of SDN and is less needing the manual IT operations and maintenance in a day-to-day routine WAN management with automating those tasks using an SD-WAN solution.

REFERENCES

1. Cox, J. H., Chung, J., Donovan, S., Ivey, J., Clark, R. J., Riley, G., & Owen, H. L. (2017). Advancing software-defined networks: A survey. *IEEE Access*, 5, 25487–25526.
2. Jammal, M., Singh, T., Shami, A., Asal, R., & Li, Y. (2014). Software defined networking: State of the art and research challenges. *Computer Networks*, 72, 74–98.
3. Jain, R. (2013). Introduction to Software Defined Networking (SDN). *Washington University in St. Louis*, 11(7), 1–44.
4. Kreutz, D., Ramos, F. M., Verissimo, P. E., Rothenberg, C. E., Azodolmolky, S., & Uhlig, S. (2014). Software-defined networking: A comprehensive survey. *Proceedings of the IEEE*, 103(1), 14–76.
5. Xia, W., Wen, Y., Foh, C. H., Niyato, D., & Xie, H. (2014). A survey on software-defined networking. *IEEE Communications Surveys & Tutorials*, 17(1), 27–51.
6. Azodolmolky, S. (2013). *Software Defined Networking with OpenFlow* (Vol. 153), Birmingham: Packt Publishing.
7. Brief, O. S. (2014). OpenFlow-enabled SDN and network functions virtualization. *Open Networking Foundation*, 17, 1–12.

12 Cloud Computing

ABBREVIATIONS

HaaS Hardware as a Service
IaaS Infrastructure as a Service
NIST National Institute of Standards and Technology
PaaS Platform as a Service
SaaS Software as a Service

12.1 INTRODUCTION

Cloud computing is a technology that allows access to its users from any device without the limitations of local networks or personalized devices, making use of the Internet to consume or save various information stored on external servers; the use of cloud computing is referred to the different services offered through the network. Today, the use of Internet services and the hosting of information in the cloud have become so relevant that companies have been forced to adapt to these new technologies, taking constant innovation processes that allow them to take advantage of these resources and make new proposals aimed at the market, since those that ignore these advantages risk being outdated and perhaps out of business.

This chapter seeks to provide conceptual bases on cloud computing and its different processes, types of service, deployments, and advantages of using cloud storage; according to [1], "Cloud is a type of parallel and distributed system that consists of a collection of interconnected and virtualized computers that are dynamically provisioned and presented as one or several unified computing resources based on data and Service Level Agreements (SLAs) established through negotiations between the service provider and consumers". In addition, it should be noted that when negotiating and choosing the cloud service provider, the service level agreements are critical since they establish responsibility and detail and clarify the quality of service. Reference [2] defines cloud computing as "a model for enabling ubiquitous, convenient, and on-demand access over the network to a set of shared and configurable resources (such as networks, servers, storage capacity, applications, and services) that can be provisioned and released with minimal management by the service provider". It is convenient to highlight "access on demand" as the main feature of cloud computing since it allows today's organizations to have resources. According to [3], "Cloud is a trend, an enabler of new business models and the machine that leads to digitization (...) Factors such as availability and scalability play an important role. Late or sooner or later, the digital transformation will affect every company in every industry". All this follows that connected resources and shared services through the Internet on demand, allowing organizations to adapt to the changing environment in a cloud computing model.

DOI: 10.1201/9781003302902-12

It has come to play an important role in the business since markets are increasingly changing and complex, so companies need to have IT and tools to adapt to this changing and complex environment successfully and quickly. According to [3], "Digitalization not only requires a transformation of the business models themselves but also of sales channels, communication with the client and partnership models. Three key trends affect the market: Greater transparency: Compare prices, product/service details, customer reviews that can be found online, Compliance with standards: Fast-moving markets and new communication concepts such as enterprise-wide collaboration systems or full deployment of virtual work solutions; and Product diversity: Competitors increase their product offerings globally through new channels such as online marketplaces. Companies have to be more flexible and agile to respond to this market". The three mentioned trends highlight the growing demand for agility, which requires highly diversified sales processes and solid and correct partnerships that generate new ideas and products for customers – increasingly demanding users. One of the best strategies to handle these customer requirements is teamwork.

12.2 BACKGROUND

John McCarthy in 1961: "If computers of the kind I have advocated become the computers of the future, computing may 1 day be organized as a public service just as the telephone system is a utilitarian public service. … Utility computing could become the foundation of a major new industry".

Although the concept of cloud computing has recently become popular, other concepts behind it have emerged even since the beginning of computing.

1955: Due to the high cost of computing in those years, John McCarthy proposed renting computing resources on a time-share basis for companies that could not afford the technology.

1969: JCR Licklider developed ARPANET; he wanted everyone in the world to be connected and access programs or data from anywhere, and Leonard Kleinrock, who seeded the Internet, stated: "So far, computer networking is still in its infancy, but as it grows and becomes more and more sophisticated, we will likely see the expansion of 'computing utilities'…".

1972: Milestone of virtualization, IBM developed the emulating VM/370 operating system.

1980–1989: PC boom, users could access remote resources via modems at 300 bits per second speeds.

1990–1998: The term "cloud computing" is first used in a presentation by Compaq Computer engineers.

1999: Salesforce emerges, a service that offers company sales information, collaborations, storage, and reports, all through a web portal.

2002: Launch of Amazon Web Services.

2006–2012: Growth of Google and its cloud-based services. Cloud storage services such as Apple iCloud (2011) and Google Drive (2012) emerge.

The rise of Google brought with its massive investment in huge server farms that would later give rise to the wide range of Web 2.0 applications and features that Google now offers as an industry titan, probably the cornerstone of what Web 2.0 is. Cloud computing today leaves software and networking companies like Novell and Microsoft in the dust, trying to find their way (Blokdijk & Menken, Brief History of Cloud Computing, 2009).

According to Joyanes (2011C), cloud computing is the great challenge that Information Technology (IT) departments have to face, which will begin to affect today's companies. As a result, managers of IT departments must consider how information is required to be obtained and distributed in shared environments in such a way as to protect organizational interests.

2012–2017: Cloud boom caused by powerful mobile devices, better networks, and faster Internet access.

2017–2022: Tools were modified to work with containers. The Kubernetes were available as an open-source product to automate application deployments, scaling, and management as a container-orchestration system designed.

The future of cloud computing in times of the coronavirus pandemic accelerated the use of automated data governance software on the Internet for working remotely, so the future of the cloud is in continuous growth (Figure 12.1).

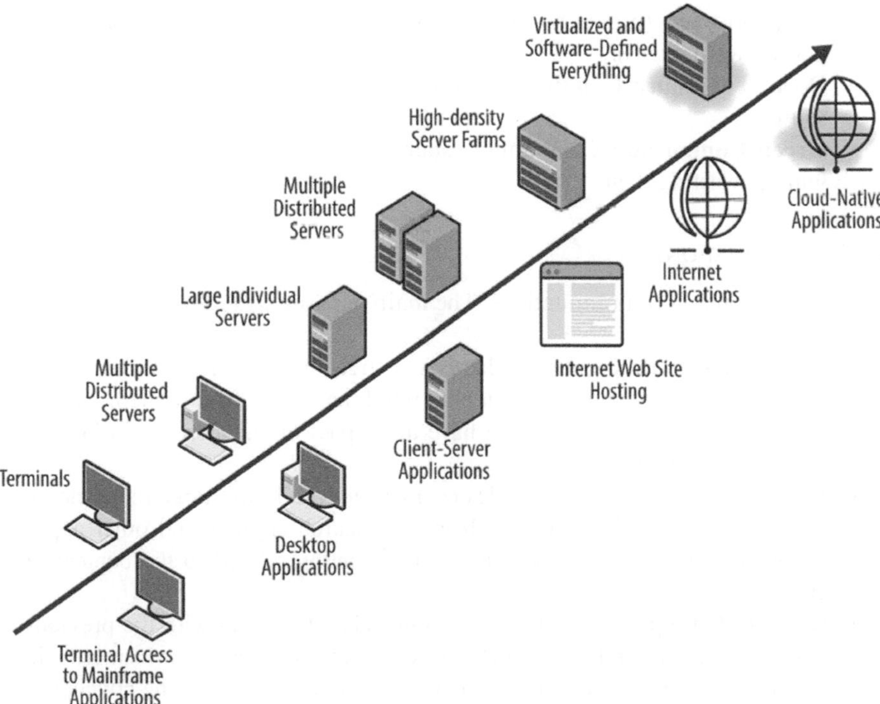

FIGURE 12.1 Evolution of cloud computing. (Note: cloud computing considers how application platforms versus computing technology have matured as the industry evolved into the era of cloud computing. Source: Bond, Planning and Architecture (2015).)

12.3 BENEFITS AND LIMITATIONS OF THE CLOUD

12.3.1 Benefits

Cloud computing provides various benefits to consumers, among which we have the following:

- **Flexibility:** Cloud-based solutions are often flexible in design: they can grow or shrink as the business requires.
- **Automatic Updates:** Consumers will not have to worry about updates, unlike manually updating servers.
- **Disaster Recovery:** Providers are often concerned with implementing robust data recovery solutions in the event of disasters.
- **Reduced Costs:** Consumers need to invest in what they will use.
- **Collaboration:** The cloud facilitates coordination and teamwork between workers in the same company that uses cloud-based solutions.
- **Security:** Some consider that having data stored in the cloud is more secure than locally.
- **Access from Anywhere:** Anyone with Internet access can access cloud services, making work easier for employees.
- **Competitiveness:** The use of cloud-based solutions improves a company's competitiveness (providing various tools of various kinds that result in competitive advantages).
- **Document Control:** In the case of joint work on documents, there will be sources.
- **Green Computing:** Computing resources are used efficiently as needed, without wasting them.

12.3.2 Limitations

The cloud also has certain limitations. The main ones are as follows:

- **Data Movement:** Currently, there is a great movement of information, either entering or leaving it, so it is intuited that in the future, there may be much more movement; this can have an impact on an increase in latency and cloud capacity.
- **Loss of Control:** A clear example occurs when emails are stored in the cloud; a bot usually examines these to help later search engines and other applications to show us advertisements or information related to the content of emails.
- **Security Perception in the Cloud:** It goes hand in hand with the previous point, in reference to the fact that it is not certain whether the information in the cloud is private, making a comparison with data centers, the concept of security as such can be reevaluated.
- **Uncertain Performance:** Although each Virtual Machine assigned to the user is isolated, it is still sharing resources with others, causing inadequate performance, such as bottlenecks, which often go unnoticed.

12.4 DEPLOYMENT MODELS

The definition of cloud is a synonym of the Internet in scientific terms, a simple network of interconnected parts of data and devices that make up the cloud. Today, public and private clouds emerge based on their Internet relationships with small, medium, and large businesses [4]. Public and private clouds are known as internal or external networks, just like corporate or cloud data centers; in practice, the difference lies in companies' relationships with the cloud.

Also known as deployment models, we have two main ones (public and private) and two more specialized ones:

- **Public Cloud:** The public cloud is a collection of hardware, networks, storage, services, applications, and interfaces owned and operated by a third party for use by other companies or individuals.
- **Private Cloud:** The private cloud is a collection of hardware, networks, storage, services, applications, and interfaces that an organization owns and operates for use by its employees, partners, or customers [5]. A third party can create and manage a private cloud for the exclusive use of another company.
- **Hybrid Cloud:** It combines the two previous ones; the main objective is to combine both services, appropriately unifying them.
- **Multicloud (Multicloud):** It occurs when an organization uses more than one public cloud (Figure 12.2).

It is necessary to emphasize that using the last two definitions will be correct as long as it generates value for the company that uses it.

12.5 SERVICE MODELS

The NIST indicates that there are three types of service models as listed below [6], and it is also possible to have other derived and specialized service models as in Figure 12.3:

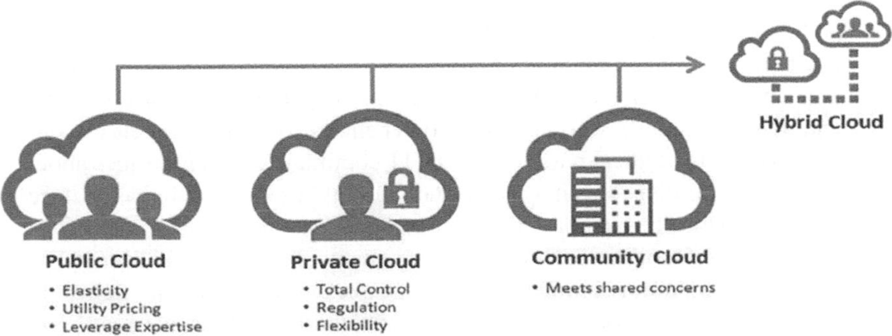

FIGURE 12.2 Graphic representation of the types of deployment in cloud computing.

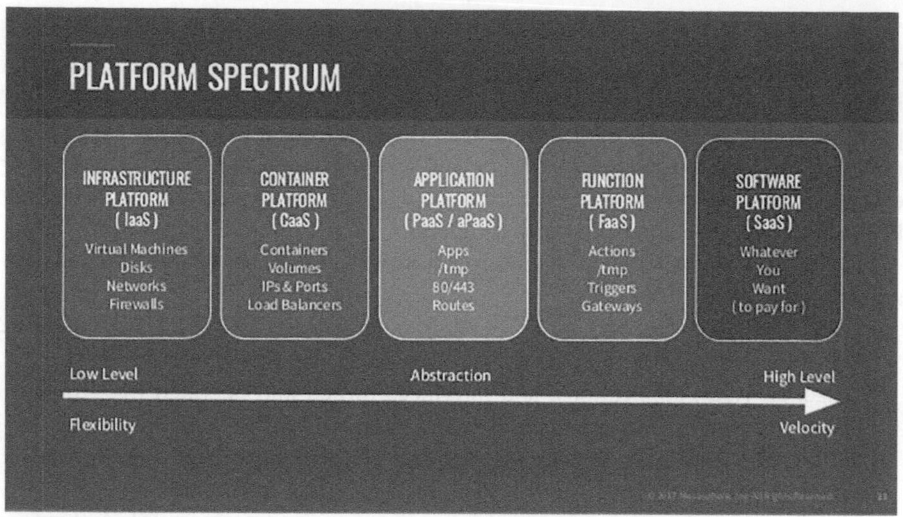

FIGURE 12.3 Cloud computing services.

- **SaaS (Software as a Service):** This is any web-based service. The infrastructure resources, among others, are the provider's responsibility – for example, the range of Google services and Dropbox.

  The concept of SaaS has been around for a long time, but perhaps in the last few years, we have clearly defined what we mean. It is any web service. We have clear examples such as Gmail, Webmail, or customer relationship management online. In this type of service, we usually access it through a browser. Moreover, all development, maintenance, updates, and backups are the responsibility of the provider. In the early 2000s, the first generation of SaaS solutions was siloed, inflexible, and designed to solve a single business problem. Since then, SaaS has evolved dramatically. Today, a modern cloud suite can span and connect everything from financial, HR, procurement, and supply chain processes to commerce, marketing, sales, and service solutions – other advantages of a modern and complete SaaS.

  Software vendors have spent the last few years attacking IT professionals and business executives with messages about the benefits of cloud computing in many ways. Some of these publications are geared toward accountants and number analysts who analyze the benefits of operating expenses over capital costs [4]. Others reach out to the IT community with messages about scalability, on-demand, and the cloud's ability to support infrastructure management tasks and allow IT talent to focus on business issues. There is a great deal of truth in each of these arguments, but very little energy has been spent explaining to managers why business applications are delivered in the cloud via the SaaS model and are paying per subscription not only makes much sense, but they are essential to closing the innovation gap that leaders often complain about in their IT departments.

SaaS is not a new concept. Web applications provided by application service providers predate the concept of "cloud computing" as we know it today. Applications initially offered through the SaaS model typically focused on sales force automation, customer relationship management, and web content management. Today, it provides a complete suite of business applications for enterprise resource planning, project portfolio management, planning and budgeting, financial reporting, human resource management, talent management, sales and marketing, customer service and support, social media, social marketing, etc. social monitoring and engagement [7]. Unlike many first-generation active server pages and other SaaS providers, SaaS enterprise applications are cutting-edge and modern, and billions of dollars are invested in developing the software and infrastructure to build and deliver your application. If a business manager is looking to gain access to the latest features without the headaches often associated with premium upgrades, he/she should explore the benefits of SaaS applications (Figure 12.4).

- **PaaS (Platform as a Service):** It provides us with the infrastructure that automatically manages the use of resources, for example, Heroku and Google App Engine. It allows the development and programming of software applications, given the low cost and the rapid opportunity offered by established channels for marketing to customers. PaaS systems are very useful as they make it easy for developers and small innovative companies to deploy web-based applications without the cost and complexity of buying servers and their corresponding configurations and implementations (Figure 12.5).

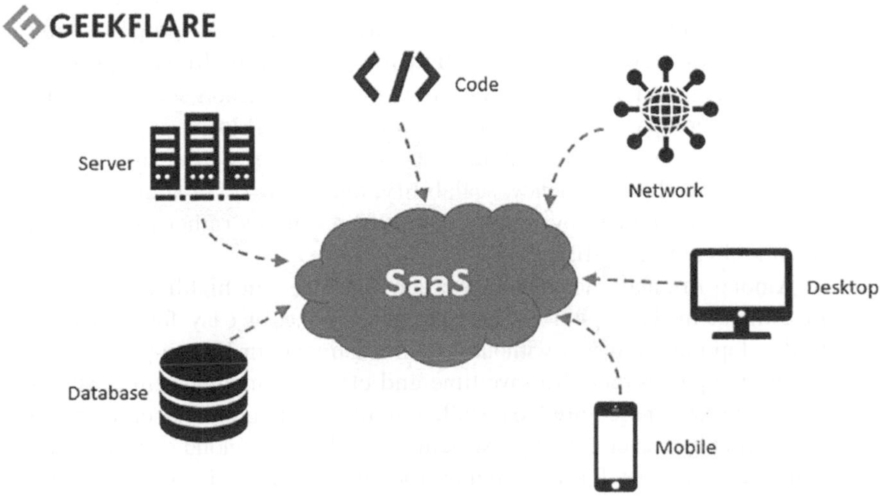

FIGURE 12.4 SaaS applications.

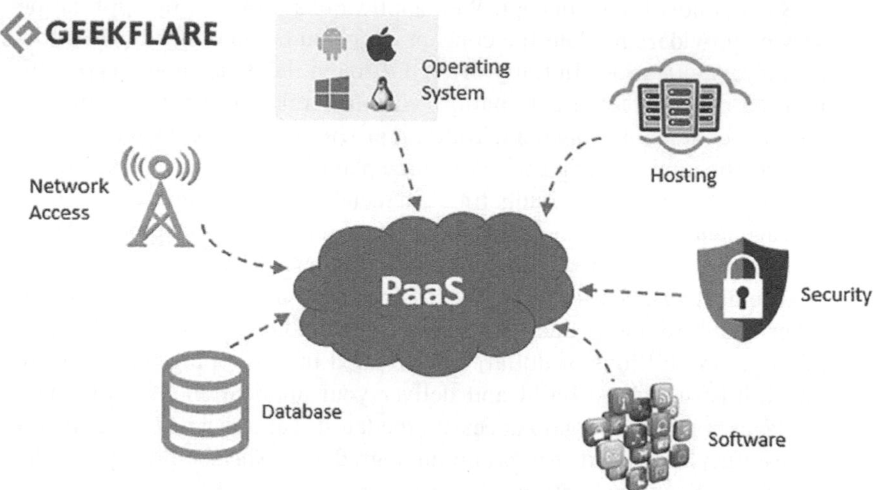

FIGURE 12.5 PaaS applications.

- **IaaS (Infrastructure as a Service):** It allows us to choose the hardware resources we will use virtually; for example, Amazon web service, with its Elastic Compute Cloud and Simple Storage Service services. IaaS, short for Infrastructure as a Service, is a cloud computing solution that includes providing and managing computing resources over the Internet such as servers, storage, network equipment, and virtualization. This computing model appeared in the early 2010s and is also known as Hardware as a Service. Since then, IaaS has become the standard model for many workloads.

  IaaS providers provide compute, storage, and networking over the Internet on demand and a pay-per-use or pay-per-use model. It is also possible to provide specialized hardware such as field-programmable gate array or graphics processing unit for artificial intelligence projects. The physical and virtualized resources offered by cloud service providers allow companies to run applications and workloads in the cloud. Therefore, IaaS is very advantageous for companies from the point of view of flexibility, efficiency, scalability, and security. In addition, users save time and money by renting compute resources rather than buying them and gaining agility.

  Among the main advantages we can find, we can highlight the lower investment in the capital because the IaaS makes it easy for companies to develop large projects without large investments in IT equipment. As a result, companies can also save time and effort. Companies eliminate the capital investment required to install, manage, and maintain an on-premises data center by outsourcing infrastructure [8]. With this cloud service model, companies only pay for the resources they need and use. IaaS providers are responsible for managing data centers containing physical machines made available to customers over the Internet, virtualized or not.

In terms of cost and investment, Infrastructure as a Service is a way to optimize your costs by delegating the administration of servers or nodes, memory, and data networks to a single operator. In some cases, the user can also delegate the virtualization layer. As it can also highlight, having more time for the user for business by leveraging the Infrastructure from provider. By implementing an IaaS platform, companies continue to maintain control over their applications, data, runtime, middleware, and operating system. They are still responsible for purchasing, configuring, and managing the software. However, the IaaS provider manages and monitors the technical infrastructure to keep things running smoothly. Plus, as mentioned above, the company's IT team does not have to worry about deploying, managing, and maintaining the physical infrastructure. Therefore, by moving from an on-premises model to an IaaS model, companies will have more time to focus on their core business while their provider's dedicated team does the best job: taking care of the infrastructure.

Finally, the IaaS model will also be able to provide the users with its maximum security and redundancy at low cost as compared to the on-premises deployment model. IaaS providers install their infrastructure in large data centers where strict redundancy and physical security measures are implemented. This will further get added to the overall security measures of the IaaS providers themselves. As a result, the level of security offered by cloud service providers will always be higher than what an in-house company can achieve. Additionally, cloud and IaaS providers often offer backup and disaster recovery solutions to help businesses protect their data [9]. This is also important in terms of cost, as companies will need to make significant investments to achieve the high availability, business continuity, and disaster resiliency offered by cloud service providers.

Furthermore, implementing a disaster recovery solution from scratch requires a significant workforce. Here is a quick guide to creating a disaster recovery plan for more details on disaster recovery planning (Figure 12.6).

From its inception, the SaaS model was designed to deliver a core set of business benefits:

We also have other derived and specialized service models, not defined by NIST but that arose from technological jargon:

- **FaaS (Function as a Service):** The type of cloud computing service allows developers to build, run, and manage application packages as functions without maintaining their infrastructure. It is an execution model that is event driven and runs in stateless containers. The functions manage the logic and the state of the servers by using the services of a FaaS provider.
- **GaaS (Gaming as a Service):** Video game services in the cloud, without the need for a console, for example, Google Stadia.
- **DaaS (Desktop as a Service):** It allows users to use their PC desktop on any device.

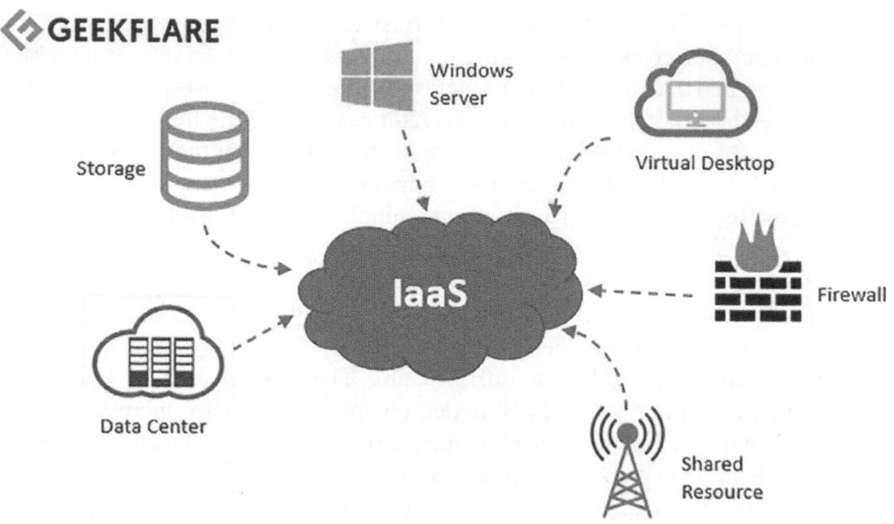

FIGURE 12.6 IaaS applications.

- **CaaS (Communications as a Service):** Cloud-based telecommunications use mobile data, such as messaging and video conferencing solutions, for example, Skype, Twitter, and Facebook. It allows users to deploy and manage applications through container-based abstraction using on-premises data centers or the cloud. The vendor provides the framework, or orchestration platform, on which containers are deployed and managed, and thanks to this organization, the most important IT functions are automated. This model is especially useful for developers, as it allows them to build scalable and more secure applications on containers. Users can purchase only the features they want (scheduling, load balancing, etc.), saving money and increasing efficiency [4].
- **DBaaS (Database as a Service):** The control and management of the database become the responsibility of the cloud service. The user only concentrates on the use of the said database, for example, MongoDB Atlas.
- **IDaaS (Identity as a Service):** Administration of identities and users based on the cloud for security, for example, biometric software or fingerprint reading.
- **STaaS (Storage as a Service):** It occurs when you buy storage space in the cloud, for example, Google Drive DropBox.
- **HaaS (Hardware as a Service):** This allows users to rent all or part of their hardware to various providers (computers, printers, telephones, etc.).

12.6 FEATURED PROVIDERS

There are a huge number of cloud service providers. Among some of the most successful and outstanding, we have the following:

- **Google:** Google, through Google Cloud, offers a series of services: Gmail, Drive, Calendar, Sites, Docs, Sheets, Slides, Hangouts, Photos, among others.
- **Microsoft:** Among the services it offers, we have OneDrive, Word Online, Excel Online, PowerPoint Online, OneNote Online, Calendar, Sway, Skype, Office 365, among others.
- drop box
- iCloud

12.7 ARCHITECTURE

NIST proposes the architectural reference model shown in Figure 12.7.

This image's most prominent and important parts are the components that make up the architecture and the services required to operate in the cloud.

12.8 CLOUD RISKS

The acquisition of technological solutions from various service providers can presume that the data is exposed (open) and dispersed outside the control of the organization. Therefore, 100% security does not exist. Susceptible information must be prioritized and controlled. Accenture, one of the world's largest consulting firms, published in early June 2010 a survey of 5,500 senior executives from nineteen countries that reveals that more than half of the world's large organizations (58%) have lost sensitive information occasionally. "Most data leaks come from within the company, usually due to employee carelessness". The main cloud computing service providers have evolved considerably in terms of security and usually offer privacy

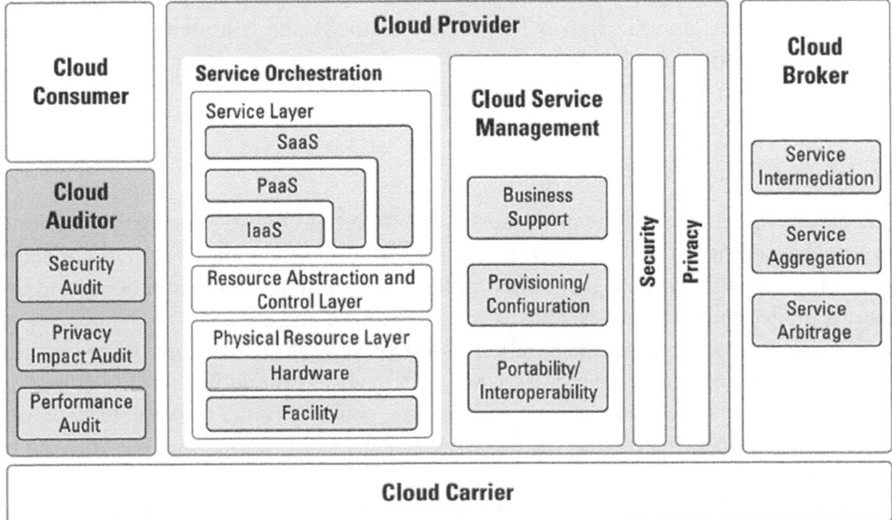

FIGURE 12.7 Cloud architectural model.

protection superior to any individual company; however, control and certification of the data are required. Internal problems are the most frequent causes of security breaches, according to the consultant above, Accenture:

- System failures or technical failures
- Incompetent employees
- Failures in business processes
- Cybercrimes
- Malicious employees
- Temporary employees or negligent contractors

12.9 DATA CENTERS AS SUPPORT FOR CLOUD COMPUTING

According to Accenture (2010), a data center is a facility used to house computer systems and associated components. This facility concentrates all or part of the resources necessary for processing information in an organization. They typically include redundant power supplies and data connections, backups, and cooling systems as safety devices. Access to these computing resources is via data connections or the Internet. In the case of cloud computing providers, these data centers also concentrate on the processing needs of customers.

Most of the big companies in technology management are expanding their data centers for their services, renting them, or subcontracting them to other companies. Microsoft is one of the representative cases of these new services. In order to compete with Google, Microsoft began to create at the end of 2007 a network of data centers, whose construction throughout the world was centered around physical surfaces of around 46,000 m^2 and costs of 500 million dollars per center.

Cloud computing is a topic in vogue, which will be growing even more during the coming years, as well as, it is observed that the Asian continent is the one that performs the most searches regarding it, a reflection of computer culture and the technological growth of that continent (Figure 12.8).

12.10 CONCLUSIONS

The cloud is a reality, and the processes and companies choose to bring this technology to their company. It is clear from the research that the cloud has many benefits and that its growth has been somewhat rapid; however, it also has limitations, and the user of the technology must be able to weigh the pros and cons.

Today, cloud computing, also called cloud or Internet computing, is among the new and best dynamic technologies since this type of computing service comes to offer a computer system as a service; in this way, end users enter a complete structure where it is possible to use of all the available services of this cloud without having to have much knowledge.

Cloud computing is thus becoming a new model for providing financial and technological services. The user has access to a wide range of services that should meet the business's needs, which will allow it to grow or shrink without problems, and in

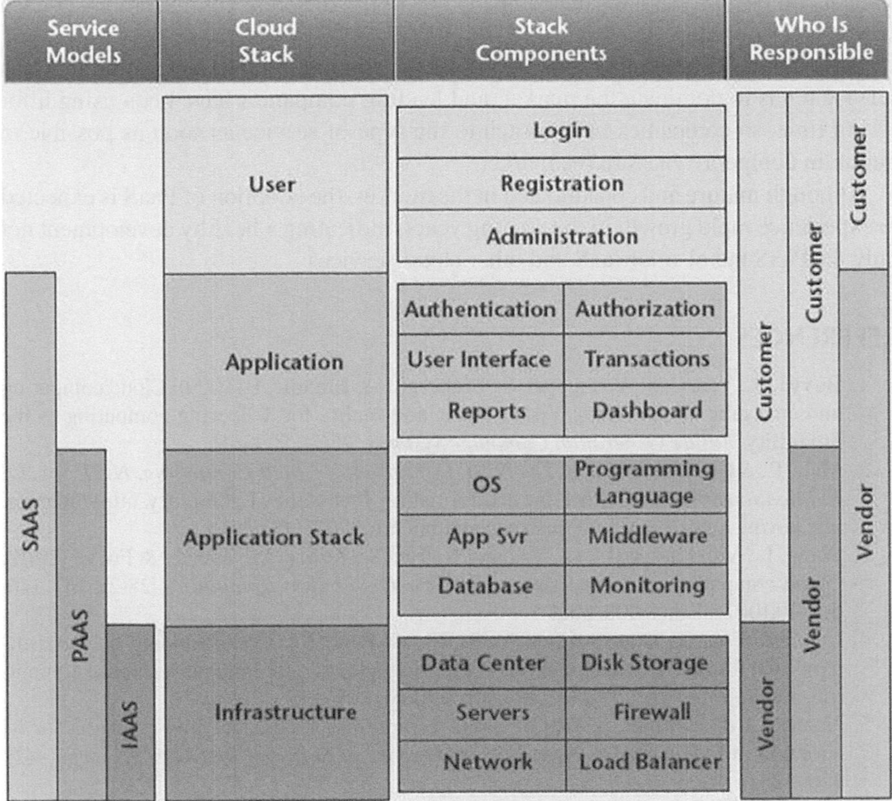

FIGURE 12.8 Cloud stack. (Note: At the bottom is the traditional data center, which may have some virtualization but does not have any of the features of cloud computing. Fountain: Kavis, 2014) [10].

the event of an unforeseen climax, we can take on the task. Ultimately, it is about making the most of server resources to provide other scalable virtual servers. The theory is good and says that it is possible to do anything in real time, but it is still a long way from having everything provided, resizing, scaling, etc.

This change in the business model benefits both providers and users. For example, the provider may have an infrastructure deployed where customers can access the services and resources they need on demand in virtual servers, virtual hubs, virtual private networks, virtual cabling, etc., and a payment model using card payment – improving in the system, time, and resources.

Companies that have adopted the outsourcing of infrastructure, applications, and services completely in the cloud or hybrid use, by sharing the use with local structures, obtain many advantages, such as reducing costs, optimizing time, and focusing on the business, increasing the return on investment.

Companies that want to enter PaaS must be aware of the migration process and understand exactly what type of service they need to navigate the nuances between IaaS and PaaS and choose the best provider in the market. The target does not change since it can be a problem for the company due to the different configurations of each server.

Although costs are falling, it is important for companies to keep in mind that this service tends to dominate the market, and leading companies have been using it for a long time, so companies must switch to the type of service as soon as possible to maintain competitiveness in the market.

Although mature and consolidated in the market, the adoption of PaaS is expected to experience rapid growth in the coming years, indicating a healthy development not only for PaaS but also for IaaS and other cloud services.

REFERENCES

1. Buyya, R., Yeo, C. S., Venugopal, S., Broberg, J. & Brandic, I. (2009). Cloud computing and emerging IT platforms: Vision, hype, and reality for delivering computing as the 5th utility. *Future Generation Computer Systems*, 25(6), 599–616.
2. Mell, P., &Grance, T. (2011). *The NIST Definition of Cloud Computing. NIST Special Publication 800-145*, Gaithersburg: Information Technology Laboratory. http://nvlpubs.nist.gov/nistpubs/Legacy/SP/nistspecialpublication800-145.pdf
3. Wang, L., Von Laszewski, G., Younge, A., He, X., Kunze, M., Tao, J., & Fu, C. (2010). Cloud computing: A perspective study. *New Generation Computing*, 28(2), 137–146. doi: 10.1007/s00354-008-0081-5
4. Lynn, T., Fox, G., Gourinovitch, A., & Rosati, P. (2020). Understanding the determinants and future challenges of cloud computing adoption for high performance computing. *Future Internet*, 12(8), 135. doi: 10.3390/FI12080135
5. Nazir, A., & Jamshed, S. (2013). Cloud computing: Challenges and concerns for its adoption in Indian SMEs. *International Journal of Software and Web Sciences*, 4(2), 120–125.
6. Makhlouf, R. (2020). Cloudy transaction costs: A dive into cloud computing economics. *Journal of Cloud Computing*, 9(1), 1–11. doi: 10.1186/s13677-019-0149-4
7. Gill, S. S, & Buyya, R. (2018). A taxonomy and future directions for sustainable cloud computing: 360 degree view. *ACM Computing Surveys (CSUR)*, 51(5), 1–33. doi: 10.1145/3241038
8. Alreshidi, A., Ahmad, A., Altamimi, A. B., Sultan, K., and Mehmood, R. (2019). Software architecture for mobile cloud computing systems. *Future Internet*, 11(11), 238. doi: 10.3390/fi11110238
9. Goyal, S. (2014). Public vs. private vs. hybrid vs. community - cloud computing: A critical review. *IJ Computer Network and Information Security*, 6(3), 20–29. doi: 10.5815/ijcnis.2014.03.03
10. Kavis, M. (2014) *Architecting The Cloud*. New York: Wiley. ISBN:9781118617618. doi: 10.1002/9781118691779

13 Internet of Things

ABBREVIATION

BLE	Bluetooth low energy
CAE	Coordination of Business Activities
CoAP	Constrained Application Protocol
DoS	Denial of Service
EIoT	Enterprise IoT
GPS	Global Positioning System
HTTP	Hypertext Transfer Protocol
IAB	Internet Architecture Committee
IETF	Internet Engineering Task Force
IoT	Internet of Things
IIoT	Industrial Internet of Things
IoMT	Internet of Medical Things
IDCI	Immersion Detection Circuit Interrupter
LoRa	Long-Range Radio
LPWA	Low-Power Wide Area
MEMS	Microelectromechanical systems
MILNET	Military Network
MQTT MQ	Telemetry Transport
M2M	Machine to Machine
NB-IoT	Narrowband-Internet of Things
NCP	Network Control Protocol
NFC	Near-field communication
ORP	Oxidation reduction potential
PPE	Personal protective equipment
RFID	Radio Frequency Identification
SNMP	Simple Network Management Protocol
USB	Universal Serial Bus
Wearcam	Wearable camera
WSN	Wireless Sensor Networks
WT	Wearable Technologies

13.1 INTRODUCTION

This technology has been transforming the way people live and work. While it still has some limitations, it offers a variety of benefits and opportunities for businesses and individuals. The Internet of Things promises to revolutionize our lives in ways beyond what we can imagine today. From smart homes to efficient factories, IoT is greatly changing the way people work and live while improving productivity and

DOI: 10.1201/9781003302902-13

profitability at the same time. Yet, despite these advances, cybersecurity remains a major concern for those adopting technology solutions like these. Smart devices have been around for a long time, but the Internet of Things is dramatically changing the way people interact with technology.

IoT allows the identification and interconnection of all objects through the Internet to connect the traditionally offline world to the online world, improve processes, increase efficiency, and reduce risks. Thus, any object with sensors and connectivity will be part of the IoT shortly, with wearable technologies being the natural bridge to achieve it. IoT will be made up of smart houses, smart cities, smart factories, and others. The first step is to provide daily life with smart objects (smart gadgets), part of which (those that can be worn) are wearable technology devices. IoT has experienced rapid expansion, with billions of devices connected with machines, appliances, sensors, wind turbines, medical devices, cars, and others.

As shown in Figure 13.1, the IoT has evolved with the convergence of microelectromechanical systems (MEMS), wireless communication technologies, and Internet microservices. They are moving from machine-to-machine (M2M) communication, machines connected through a network without human interaction, to becoming a sensor network of billions of smart devices connecting people, systems, and other applications to collect and share data. As a result, the large-scale implementation of IoT devices promises to transform many aspects of our lives.

For consumers, new IoT products such as home appliances, home automation components, and Internet-enabled energy management devices are leading us toward a vision of the "smart home" that offers greater security and energy efficiency. In addition, other personal IoT devices, including wearable devices for monitoring and managing physical activity and Internet-enabled medical devices, are transforming the way health services are delivered.

The history of the IoT continues to be written day by day. It is in full swing, with the appearance of new devices, protocols, access technologies, etc., that converge with advances in other technologies such as cloud computing, big data, and artificial intelligence, enriching and giving the IoT universe more opportunities for growth every day [1].

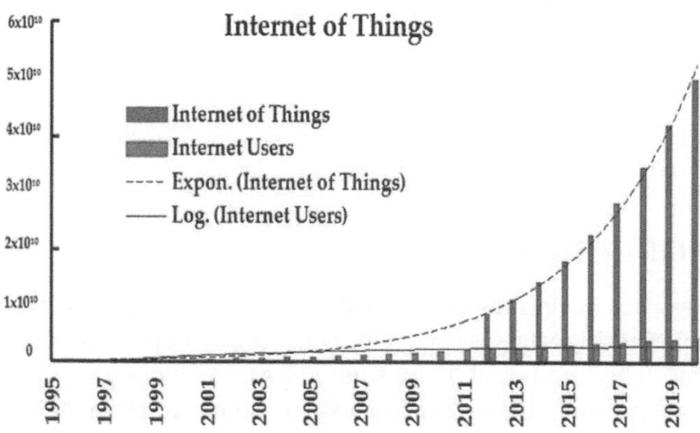

FIGURE 13.1 Growth of the Internet of Things.

13.2 IoT WORLD BACKGROUND

Despite its late conceptualization, the first cases of connected elements were found at the end of the 19th century. For example, it was in 1874 when, based on short-wave radio links, it was possible to connect a meteorological station on the top of Mont Blanc with a laboratory in Paris.

1926: Nikola Tesla, a visionary, was the prototype "mad scientist". Even though his work in electromagnetism and electromechanical engineering has resulted in advances as "palpable" for the general public as alternating current, the electric motor, radio, remote control, fluorescent lights, or low-consumption light bulbs, his controversial character and extravagant ideas were the cause of his being ostracized [2]. However, it cannot be denied that he was a true visionary in many fields. In 1926, he was interviewed by Colliers magazine, where he amazingly anticipated not only the growth of global connectivity and technological miniaturization but also the interconnection of everything in what he called a "big brain".

"When wireless is fully developed, the entire planet will become one big brain, which it already is, with all things being particles of a real, rhythmic whole...and the instruments we will use for them will be incredibly simple compared to our current phones. A man will be able to carry one in his pocket" (Nikola Tesla, 1926).

1950: Alan Turing, considered one of the fathers of computer science and the forerunner of modern computing, was another man ahead of his time. Able to see possibilities that the immaturity of the technique of his time made unfeasible until several decades later, he advanced the future need to provide intelligence and communication capabilities to sensor devices. "...it can also be argued that it is better to provide the machine with the best sensing organs money can buy, and then teach it to understand and speak English. This process will follow the normal learning process of a child" (Alan Turing, 1950, *Computing Machinery and Intelligence in the Oxford Mind Journal*).

Years 60, 70, and 80: ARPANET and first networks, in the 1960s and, above all, the 1970s, the first communications protocols were created to define the basis of what the Internet is today.

In 1969, the first message was sent through the ARPANET, the US Department of Defense's operational network, the origin of the global Internet. However, the lack of fast and low-cost communications over medium and long distances created heterogeneous networks, totally incompatible with each other. This resulted in an ecosystem of silos of locally connected equipment, typically limited to academia and the military.

1982: First "connected object" to ARPANET, in 1983 TCP/IP became the heart of the ARPANET network, completely replacing the NCP protocol, and the network was split into two parts, MILNET for military use and ARPANET for the rest.

In the 1970s, the Computer Science Department at Carnegie Mellon connected a Coca-Cola machine to the departmental server via a series of microswitches. Before "taking a walk" to the drinks machine, the interested party could check from his computer if any soft drinks were left and if they had the right temperature, knowing how long they had been cooling in the machine.

As we can see on the department's website, which summarizes the story, anyone connected to the Internet (then ARPANET) could find out the status of the machine with a "finger coke@cmua".

These programs continued to be used and updated for more than a decade (as we can see in Figure 6 of the CMU SCS Coke Machine page that lists the different generations of the software) until, in the early 1980s, the format of the bottles changed and the machine had to be changed. However, it was not until the 1990s that the students regained interest and connected the new machine to the Internet.

The 90s: the Internet revolution, in 1990: A toaster, the first "connected object" to the Internet, John Romkey's connected toaster appeared, considered the first IoT device. The connectivity was through the TCP/IP protocol, and the control was carried out through SNMP (Simple Network Management Protocol), a network management protocol used to control the turning on and off of the appliance. They could control it on, off, and "toasting" time from any computer connected to the web. The only human interaction required was…putting on the toast. The following year, however, they incorporated a small robotic arm that fully automated the process.

1993: Xcoffee project, first connected webcam. Again, with caffeine as the protagonist, in 1993, the Xcoffee project emerged. Cambridge University students developed the first online camera to monitor whether coffee was in the department's coffee machines. The original webcam updated the image of a coffee maker about three times a minute, so all concerned would be aware of when they could enjoy a fresh cup of coffee.

1994: First connected wearable camera (Wearcam), a year later, Steve Mann (Stanford University), known as "The Father of Wearable Computing" (The Father of Wearable Computing), connected the first portable camera to the web.

In the 2000s, at the beginning of the 21st century, thanks to the popularization of wireless connectivity (cellular or Wi-Fi), the first explosion in the growth of connected objects took place. This growth has been consolidated especially in recent years, as new concepts such as WSN (Wireless Sensor Networks) or new radio access technologies such as LPWA (NB-IoT, LTE-M, etc.) have emerged, to give step to the IoT that we all know finally [3].

2000: Digital Internet GOD, LG launches the first Internet-connected refrigerator. However, it was not well received since its price was very high.

2005: Nabaztag, the first connected pet-virtual assistant, the French company Violet launched Nabaztag (in Armenian) on the market. It is a bunny-shaped device that connects to the Internet via Wi-Fi waves. It communicates with the user through voice messages and changes in color or movement (of its ears!). As a good virtual pet, Nabaztag reproduces, speaks, listens, and responds to the users' voices. You can also wake them up tomorrow with current news from digital newspapers, music from their favorite station, weather information, or notify them when an email or a message arrives on their social networks.

In 2008: More connected devices than people, 2009: The term "Internet of Things" (IoT) emerged, and Google begins work on the autonomous car project. Google started its self-driving car project, the Google self-driving car project, later known as Waymo. The technology developed by Waymo allows a car to drive autonomously in the city and on the highway, detecting other vehicles, traffic signs, pedestrians, etc.

The first cardiac implant monitored by IoT 2009, Saint Jude Medical Company manufactures the first connected cardiac implants. A wireless USB adapter received the data from the implant and later transmitted it to the mobile phones of the medical staff.

2010: The NEST company, starts manufacturing smart home appliances. The first was a thermostat that optimized the heating schedule based on user usage patterns.

2016: First IoT Malware: MIRA, in 2016, Mirai emerged, a botnet targeting IoT devices, mainly routers, digital video recorders, and IP surveillance cameras. This malware collects default passwords set by device manufacturers that users often forget to change. It then uses the devices to perform Denial of Service (DoS) attacks on third parties, typically very popular web pages.

2017: IoT Services, Big cloud service providers offer IoT solutions: Azure IoT Edge, AWS IoT, and Google Cloud IoT core.

13.2.1 Communication Models of IoT

It is useful to think about how IoT devices connect and communicate in their communication models from an operational standpoint. In March 2015, the Internet Architecture Committee (IAB) released a document to guide the creation of smart object networks (RFC 7452), which describes a framework of four common communication models used by IoT devices. The following discussion introduces this framework and explains the main features of each model.

13.2.2 Communications Device to Device

The device-to-device communication model represents two or more devices connecting and communicating directly, not through an intermediary application server. These devices communicate over many networks, including IP networks or the Internet. However, protocols such as Bluetooth, Z-Wave, or Zigbee are often used to establish direct device-to-device communications.

These device-to-device networks allow devices that adhere to a certain communication protocol to communicate and exchange messages to achieve their function. Typically, this communication model is used in applications such as home automation systems, which typically use small data packets for communication between devices with relatively low requirements in terms of the transmission rate.

Residential IoT devices – light bulbs, switches, thermostats, and locks – typically send small amounts of information (for example, a door lock status message or a command to turn on a light) in an automation home scenario. This device-to-device communication approach illustrates many of the interoperability challenges discussed later in this chapter. As described in an *IETF Journal* article, "often these devices are directly related, typically have built-in security and trust [mechanisms]; furthermore, they use device-specific data models that require redundant development efforts [by device manufacturers]". This means that manufacturers must invest in developing ways to implement device-specific data formats rather than open methods that allow standard data formats.

From the users' point of view, this means that the underlying device-to-device communication protocols are not compatible, forcing them to select a family of devices that use a common protocol. For example, the family of devices that use the Z-Wave protocol is not natively compatible with the Zigbee family. While these incompatibilities limit users' choice of devices in a given protocol family, users also know that products in a given family tend to communicate well.

13.3 INTERCOMMUNICATION BETWEEN THINGS

The digitalization of intelligent machines was carried out thanks to the communication between physical equipment such as hardware and digital systems such as software, commonly called IoT. This ends up being real and very prosperous because it is projected that by the year 2023 onwards, there are expected to be 14 billion IoT devices and connected systems in the world.

Next, we can see a small standard architecture of IoT communication (see Figure 13.2). Starting from the perception layer, the stage in which the devices will be stimulated through the sensors to send information through the specific network,

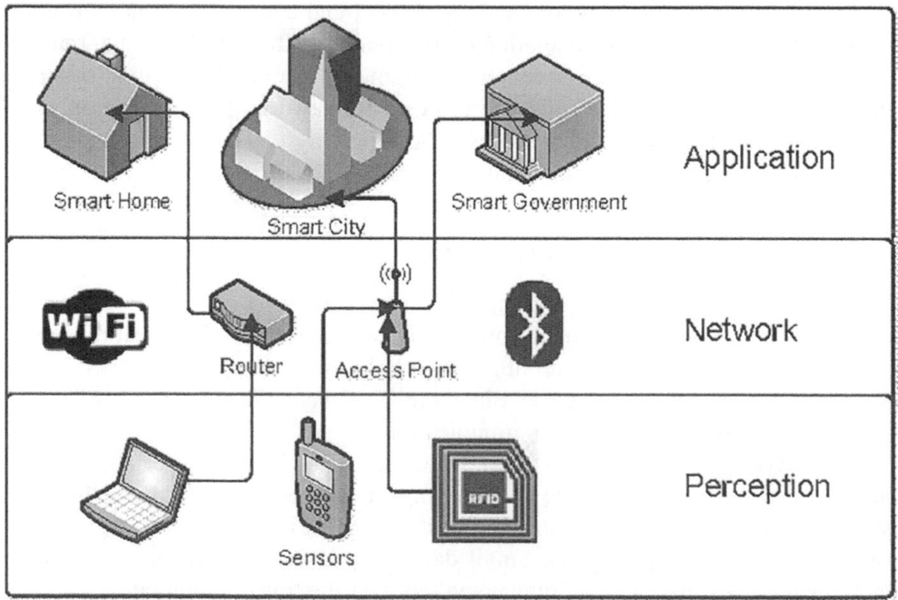

FIGURE 13.2 IoT communication architecture.

which could be sent, from a Router, an Access Point, or via a Bluetooth connection [4]. Finally, this information, which is transmitted by the devices through their sensors and channels, will land on the application that will manage, transform, and refine the information obtained to obtain results or return the information to the devices.

13.3.1 Protocols Used to Transmit Data

After seeing the flow of intercommunication between IoT devices, it is necessary to emphasize that the IoT device that communicates with the application through channels has a type of communication that uses a specific protocol that allows it to transmit information to a receiving application with well-known protocols such as HTTP, MQTT, and CoAP.

13.3.2 Embedded Vision Systems

We know an embedded system as an electronic device with computational intelligence designed to fulfill one or several related tasks determined from the design and, therefore, are predictable when executed in real time, in addition to being made up of components of hardware and software.

These generate a great benefit in IoT because the sensors already contain an integrated system that facilitates communication between sensors, providing information to the device through digital processing and intelligent algorithms that interpret the meaning of images or videos. In addition, they visualize what is happening and what needs to be done. This leads to the appearance of very powerful, low-cost, and low-power processors.

13.3.3 Webinars on Embedded/Integrated Vision Systems

One of the characteristics that represent IoT is the use of sensors. These have an integrated system, which allows them to facilitate and divide the processing logic from the same sensor by sending responses to the central system. Many sensors are part of IoT devices, such as proximity, infrared motion, temperature, stepper motors, and vision sensors. The importance of sensors is that there are various conferences where different sensors are discussed. In this case, the vision sensors brought together thousands of professionals in the area over the years. Some of them are listed below.

13.3.4 Machine Learning

From its origins, it is known that IoT is linked to machine learning, just as human learning is based on past experiences. Analogously, since the system learns from previous situations, added to this and handling large masses of data allows the devices to have constant learning.

13.3.5 Arduino

Arduino is a free software and hardware development company with one of the largest development communities in the world. For the most part, it is in charge of the design and manufacture of hardware development boards to build digital devices that interact, detect, and control real-world objects, strongly supporting the IoT.

They are very commercial in the market; due to this, there are different types of Arduino according to the needs and scope of the system to be implemented. Next, the most sold types of Arduino in the market are detailed: Arduino UNO, Arduino Leonardo, and Arduino 101 (Figures 13.3–13.5).

Each type of Arduino was developed for specific purposes. Therefore, it cannot be determined if one Arduino is better than another since each one is the best in its field of action. Even so, it is possible to rescue some differences between the three best-selling Arduino boards on the market (see Table 13.1).

FIGURE 13.3 Arduino UNO.

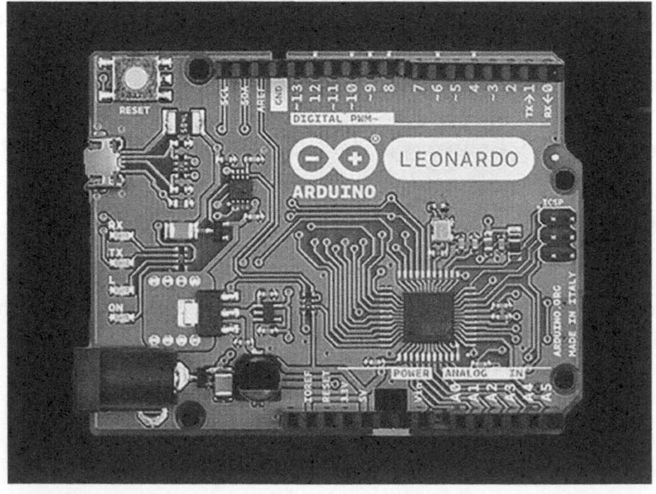

FIGURE 13.4 Arduino Leonardo.

FIGURE 13.5 Arduino 101.

TABLE 13.1
Comparison of the Three Best-Selling Arduino

Characteristics	Arduino Uno Rev 3	Arduino Leonardo	Arduino 101
Microcontroller	ATmega328P	ATmega32u4	Intel Curie
Voltage (V)	5	5	3.3 (5V tolerant I/O)
Input voltage (recommended)	7–12	7–12	7–12
Input voltage (limit)	6–20	6–20	7–17
Digital (I/O) pins	14 (4 for PWM output)	20	14 (4 for PWM output)
PWM digital (I/O) pins	6	7	4
Analog input pins	6	12	6
Analog input channels	–	–	4 of the digital I/O pins
Bluetooth	–	–	Bluetooth LE
Additional characteristics	–	–	6-Axis Gyro/Accelerometer
Length (mm)	68.6	68.6	68.6
Width (mm)	53.4	53.3	53.4
Weight (g)	25	20	34

On the other hand, the Internet of Things was, is, and will always be a fundamental factor in technological and social development due to the great benefits it offers for the implementation of new devices and the improvement of the quality of life of society. One of the primary keys to the success of IoT is due to the large community of developers who are aware of new updates, due to the great variety of new sensors that are developed daily and the new needs that arise in the global context.

Finally, IoT is still observed not used much in agriculture. Therefore, there is a need to improve IoT applications in this field and educate farmers, reducing human dependency and improving yields, leading to an increase in the economy.

13.3.6 The Internet of Things as an Ally of Digital Transformation

The Internet of Things is growing and gaining strength in fields ranging from automation and robotics, industrial and agro-industrial processes, to an open approach to solving various societal problems; in this multifield environment, the IoT becomes important as a link and route to the world's connectivity. Being connected to the world will allow us to be competent and analyze our data and that of the world in real time. According to Peter Newman in the IoT Report: How the Internet of Things technology growth is reaching mainstream companies and consumers, "The continued growth of the IoT industry will be a transformative force in all organizations. By integrating all of our modern devices with Internet connectivity, the IoT market is on track to grow to more than $3 trillion annually by 2026" [2].

13.4 INTERNET OF THINGS 2.0: THE NEXT STEP TOWARD INDUSTRY 4.0

Talk about the Internet of Things is usually referred to as connectivity: devices, connections, and data volumes. However, the new label of the Internet of Things 2.0 also implies a transformation process. This transformation is technological and focuses on the integration of all available elements. Therefore, 2.0 may be the later stage of the IoT when its previous version is as established as today's Internet or electricity. At that time, the focus will be on the possibilities of improving business, life, or society thanks to the knowledge acquired and hyperconnection.

To join this, each company must create a roadmap with specific objectives that take real life and the massive opportunities of this paradigm shift. Hand in hand with the inclusion of the Internet, robotics, big data, artificial intelligence, or the Internet of Things itself in the industrial sector, we are moving toward what is known as Industry 4.0 or the Fourth Industrial Revolution. The merger between Industry 4.0 and the Internet of Things will bring numerous benefits to companies, especially considering that the crisis derived from the coronavirus is driving change in the industrial sector. The benefits that a company can obtain by implementing the Internet of Things:

- Lower production costs.
- Increased productivity by streamlining production deadlines.
- Better levels of quality and efficiency.
- Better response and flexibility in the face of possible errors in the production chain.
- Greater ability to personalize products and services.

Contribution of added value thanks to intelligent manufacturing that knows how to respond to the needs of a connected society.

In short, it is expected that the Internet of Things will be consolidated in various fields, including companies and industries, which will open up a whole range of opportunities in the coming years.

13.5 IoT APPLICATIONS

To understand what IoT is, it can be helpful to know how it can be used. Being familiar with some aspects of IoT is important, as the concept of a smart home or connected home comes from the term IoT. At the most basic and understandable level, we have meters for gas and electricity. These little devices have been around for several years, and their job is to monitor energy usage, communicating the data directly to the energy provider. Contrary to the old technologies, there are a number of benefits with this technology. First, the supplier can obtain more information about when and how their energy is used and trust the accurate report. The consumer or end user can see exactly how much energy they are using, giving them the control to regulate their consumption.

However, the smart home can be much more connected. The biggest attraction of IoT is that it allows access to the network from anywhere. This means it is possible to access the home network from any location remotely. When it comes to energy, we can control using certain devices that, for example, allow us to turn on the heating before we get home. Other benefits include turning lights on and off when users are away from home and interacting with their pets thousands of miles away.

However, IoT is not just a technology for home use. Companies are investing in the great benefits it can bring them. Courier companies, for example, are using RIFD chips in high-value packages to scan them as they enter and leave and even monitor information such as how roughly the package has been handled. The dispensing machines can make their product orders in real time to the warehouses. Even animals can carry devices that can monitor their location or health. Machine-to-Machine (M2M) communication is part of IoT, and you can see some examples of how it works by accessing this blog here.

13.5.1 What is IoT, and what are its Main Applications

There is much talk about the Internet of Things or the Internet of Things. However, many people still do not know what IoT is or what it is for. It is estimated that the IoT already connects 31,000 million devices worldwide, but many of us live oblivious to an interconnected reality that will change all the rules silently, ubiquitously, and without brake. So is the IoT.

13.5.2 This is What a Home IoT Network Looks Like

In the first example, we look at a home IoT. A home with the router as the center of the network from which a smart TV, the family's different smartphones, the home computer, and a small network of smart light bulbs are "hanged", all connected via Wi-Fi.

In addition, the same network has Bluetooth subnets, such as the one with the television with the remote control and the refrigerator and the different telephones with the robot vacuum cleaner, the door lock, or some activity bracelet.

The possibilities are countless, from devices for private use such as virtual assistants from Amazon or Google to real-time monitoring systems. Whether the sector or the type of industry, today's IoT solutions can be applied across the board. Smart devices can provide greater efficiency, personalization, comfort, improved production, and safety.

13.5.3 Business

The term "Enterprise IoT" (EIoT) refers to all devices in the business and corporate environment. Enterprise IoT solutions enable them to improve current business models and build new relationships with customers and partners. However, its implementation presents certain challenges. The volume of data generated by a system of smart devices can become overwhelming. Integrating big data into existing systems and setting up analytics to use that information can be challenging.

In addition, security is a critical aspect when designing IoT systems. Still, many companies find implementing this technology worthwhile; therefore, it is possible to find successful case studies in almost all sectors [5].

13.5.3.1 Hostelry

IoT devices are already being used in this sector. Surely you have already seen how some fast food restaurants have a terminal that notifies the client when the food is ready. Even the waiters can locate the table where the food is going.

If we look from the management side, food stores have a lot to say. For example, it prevents food expiration, needs, automatic orders, and management of devices such as cold rooms, kitchens, and ovens.

The company Powerhouse Dynamics offers an IoT application for this sector. It allows to control, supervise, and manage centralized air-conditioning equipment, control lighting, supervise refrigeration equipment, detection and misuse of water, food storage monitoring, and maintenance monitoring.

13.5.3.2 Business IoT Network

In the second IoT network, in an Internet of Things network at the enterprise building level, in this case, the company network connects several routers and systems in the same building. The IoT office is born. These systems, in turn, are nodes or central points of smaller systems. This branching is called "hierarchy" and is very common in all computer networks.

13.5.4 Vehicle Fleets for Logistics

Logistics is one of the sectors where IoT applications have the most influence. Control packages, managing vehicles, preventing theft, and movement management are some aspects to consider within this area. At this point, the technology is very advanced. For example, some services allow people to see the location of the package in real time.

13.5.5 IoT Applications for Home Use

It is being able to create IoT applications to use in the home. It is possible to see in which branches the development of IoT applications is most advanced. Home automation: One of the significant challenges of technology is home automation. It has not yet been able to take off due to the difficulty in the infrastructure and the absence of standards that allow communication between devices of different brands.

13.5.6 Growth of the Number of Connected Devices

Currently, studies carried out by large multinationals in the IT world predict that this great brain will be formed, in 2020, by some 30 billion Americans of "intelligent things" connected to the Internet and, just 4 years later, that figure will double. This exponential increase, as seen in Figure 13.6 in the number of elements connected to the Internet, means that implementing IoT technologies in the world in which we live offers society a wide range of new services and innovation opportunities (Figure 13.7).

13.5.7 IoT in Agriculture: Smart Farming

Smart Farming was a pilot experience of agriculture and livestock, was a highly unpredictable casuistry due to its high correlation with the weather and environmental conditions. For this reason, the use of technology will reduce the human factor in the supervision of plantations and increase productivity and profitability by controlling the state in which they are found easily. To do this, the automation and sensorization of parameters that affect it, such as the weather, the state of the air, the land, or even the state of health of the animals, is implemented on the farm. This information

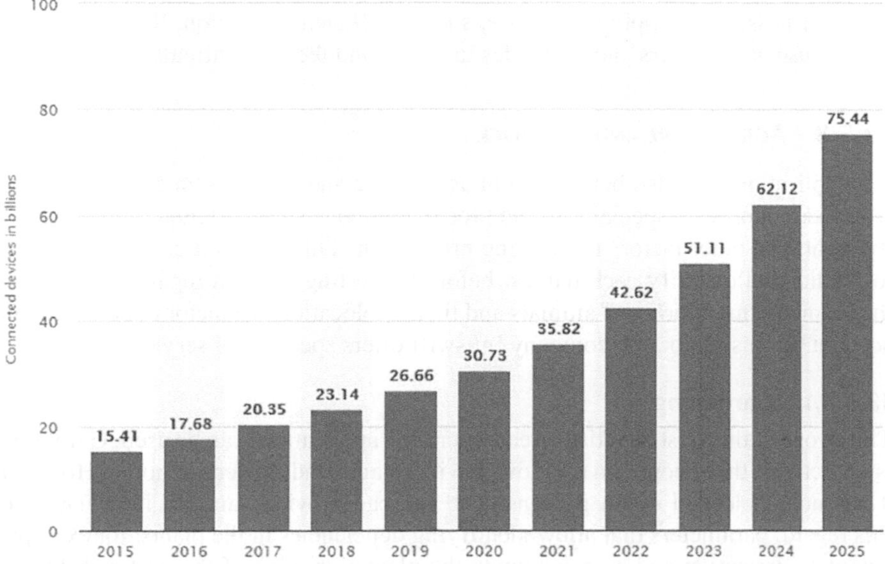

FIGURE 13.6 Exponential increase of elements connected to the Internet that implementing IoT technologies in the world.

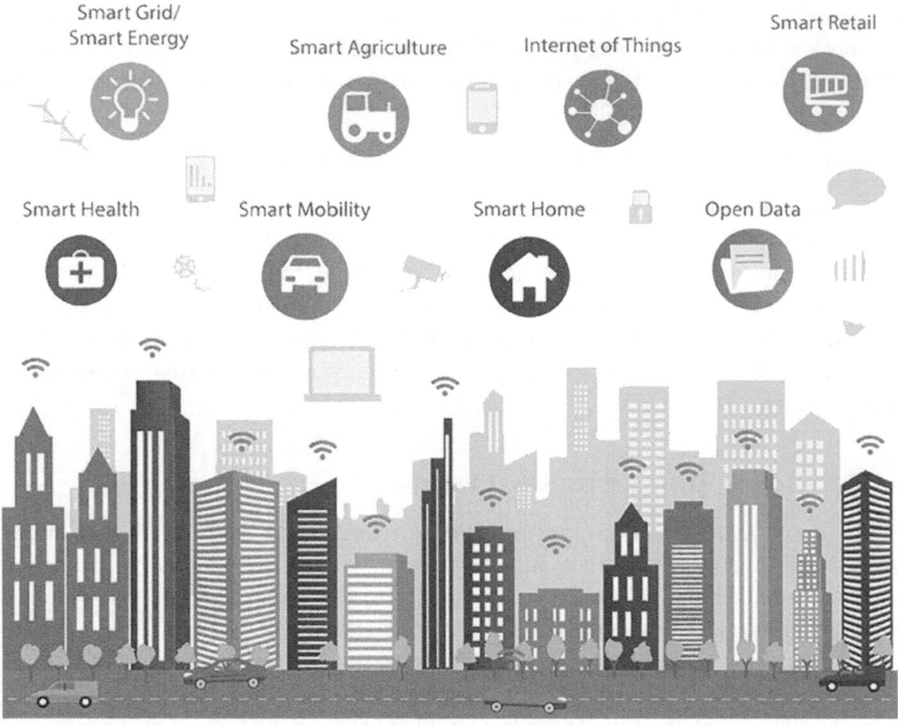

FIGURE 13.7 Examples of the new smart era.

makes it possible to apply techniques, such as efficient irrigation, the specific and precise use of fertilizers and pesticides in crops, and feed for animals.

13.5.8 Agriculture and Livestock

IoT applications are also being used in agriculture and livestock in agricultural and livestock countries, especially to monitor magnitudes such as temperature, humidity, light, and other factors influencing production. This makes it easier for farmers to predict and quantify each harvest before harvesting it. With regard to livestock, the biometric monitoring of animals and their geolocation is a factor to be taken into account. In this sector, the company Infiswift offers specialized services.

13.5.8.1 Gardening

This is one of the most beautiful areas where IoT applications can be used. Gardening is an activity that people do and love having plants and flowers at home. However, it has not yet leaped urban gardens. The IoT can provide valuable information in this regard, parameters that allow identifying deficiencies in the plants, for example, automatic irrigation and even automate the plantation. One of the best well-known IoT apps is Farmbot. The best thing is to watch the video to understand what it can do (Figure 13.8).

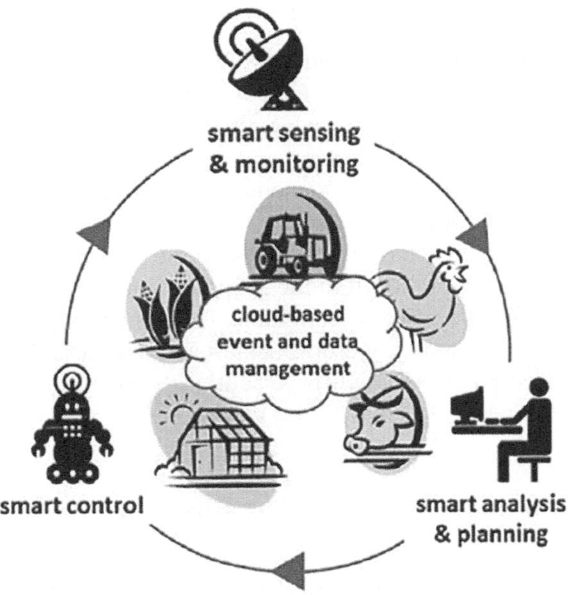

FIGURE 13.8 IoT applications getting used in agriculture and livestock.

13.5.9 IoT in Medicine IoMT: Smart Health

Everything related to the world of medicine has as its main purpose to offer the highest possible quality in patient care, diagnosis and intervention. Reducing long queues, inventory control, or greater control of the status of patients, both inside and outside the hospital, are some of the possibilities that the application of IoT solutions to medicine can make a reality. As it is an aspect of the utmost importance, it will never seek to replace the human being, but instead will seek to save time for the medical staff and more complete and personalized attention.

To achieve these objectives, it is again necessary to implement an extensive network of sensors, among which wearables can be included, which monitor the status of hospital patients or those who have been discharged and are prone to need rapid action (telemedicine). In the same way, if we extend this network of sensors to the real-time location of both the hospital's medical staff and the medical teams, and we coordinate everything efficiently, we will be able to reduce the reaction time in an emergency proportionally to this efficiency.

13.5.9.1 Health

This issue concerns us all since, without health, the rest of things do not matter. IoT applications are used for telemedicine, real-time patient monitoring, early diagnosis, and others. In this sense, the Spanish company Libelium has a complete device that helps us monitor patients. It is based on Arduino and allows remote monitoring of vital signs. As Libelium's IoT application is called, it is useful in remote areas with a shortage of doctors.

13.5.10 IoT in Energy Management: Smart Energy

Together with the Smart Industry (or Industry 4.0), IoT solutions in the energy world form a tandem with great possibilities of becoming the field with the highest implementation rate in the coming years. Today, these solutions are mainly focused on improving the comfort situation of users: management of temperature, lighting, or even appliances. In this case, the network comprises a network of sensors from which to extract current conditions and of devices that, through their connectivity, allow us remote or programmed control of their operation. Similarly, the efficient management of these resources is added to the scope of Smart Energy, with the aim of more economical and sustainable consumption. The primary objective is to increase energy efficiency without reducing user satisfaction. To achieve this, the processing of all the sensorized parameters will be essential to establish an efficient control logic that will achieve this objective.

13.5.11 IIoT, the Industrial Internet of Things

In addition to the above, one of the most widespread applications of the IoT is the Industrial Internet of Things (IIoT) as well as the immersion detection circuit interrupter (IDCI). It is widely used in all kinds of industries:

Logistics: For example, to know where each truck, container, and object on the distribution line is. They are also very useful at the warehouse level. Both distribution companies, large airports, and even large stores monitor objects and robots within their facilities to optimize routes, lower costs, or add security.

Manufacturing: The Internet-connected robots we see on a car assembly line are IIoT. So are the serial printers of a publisher or the cutting elements of a sawmill if they are connected to the network. Every manufacturing process can be connected and become part of the Internet of Things, from PET bottles to stuffed animals.

Construction: It is increasingly common to have work machinery connected and sensorized. Knowing where a backhoe is important, but analyzing its fuel, state of conservation, or having each component digitized in real time adds a lot of value. In addition, helmets and other IoT PPE are already being used, and warnings are issued in the event of occupational accidents.

13.6 ADVANTAGES OF IoT

What does the implementation of IoT systems look for in EPIS, work teams, and the environment?

The installation of sensors and IoT on different media enables certain end uses: depending on the configuration applied to them, they will capture one type of data or another and thereby provide specific end uses. Agility in the management of EPIS. This technology provides digitized information on the PPE and enables their traceability; in this way, it is possible to know the characteristics of each PPE, following its acquisition, purchase, and subsequent delivery to users, in addition to controlling maintenance and its times, stock and track expiration and withdrawal. Automation of monitoring the use of EPIS. Guarantee the control of staff access to certain areas. For example, control access to an area or a machine verifies that the profiles of people are appropriate in each area, with the appropriate training and PPE. This applies both to internal personnel and to those who come from third-party companies but attend the same work center, thus facilitating the management of the Coordination of Business Activities (CAE). Based on the environmental working conditions and thanks to the IoT systems, it can be verified that the PPE fulfills the function for which they have been foreseen and the interaction with other equipment and machines, give the possibility of turning machines on and off depending on the user and connecting and avoiding vehicles.

Reduction of response times in the event of an accident, is important with real-time detection of a dangerous situation allows protocols to be activated instantly. Geolocation in emergencies improves the effectiveness of the intervention since it allows:

- Start the work of locating and searching for people from the beginning of the alert.
- Know the emergency and notify the relevant rescue team based on its characteristics.
- Notify the pertinent rescue team based on its characteristics, adapting it to the magnitude of the danger.
- Geolocation, alarm, and data management system speed up and improve the response to emergencies.
- Continuous data collection. Unlike the currently carried out measurements, they allow alerts to be established based on measurement values and action management.
- Improvement in accident investigation. More significant data is available in the phases prior to the event.
- Predictive prevention anticipates and predicts an IoT platform collects and processes large amounts of data from multiple devices and analyzes it to establish patterns.

From the data of the working conditions (metrics of different parameters), together with those of other sources, you can establish the probability of an accident occurring in a specific situation and generate alerts, thus helping to define preventive actions.

13.7 LIMITATIONS OF IoT

13.7.1 Considerations to Integrate IoT Technologies

To develop an adequate project for integrating technologies in PRL, it is necessary to plan and consider some points before undertaking any action. First, think in the long term why the IoT system will be implemented and the purpose of collecting the information to avoid leading to an excess of information that is not useful. It is also important to think about the company's digitization strategy, to choose compatible technologies in the long term, and assess the possibilities and services they offer in the future. Define the needs in IoT: what variables to measure and the objective. At a technical level, it is convenient to study before implementation the typology of sensors that exist in the market and their characteristics and the communication networks necessary for their connection with the platform that collects the data. In this part, it is crucial to look for a platform that offers data processing solutions that fit its perspectives. Carry out tests on the ground, especially regarding information communication, to identify possible interference or lack of signal. Verification of the installation is essential to select the correct communication protocol. This can be a limitation on some occasions. Finally, consider the ownership of the data when negotiating with suppliers, assess and be clear about who owns the data, and what limits of exploitation exist. Who owns this property, the company that generates them, or the one that collects them? To avoid problems in this regard, platforms that guarantee data ownership and protection must be used.

Take into account the following factors that can be limiting at the time of project implementation: Communication with the different sensors to be able to collect the data that is being obtained, avoiding "dark areas" that can be the companies' infrastructures (machines and installations) that make connection difficult. Connectivity: Through the different communication technologies: Wi-Fi, Bluetooth/BLE, NFC, RFID, GPS, Zigbee/Thread, LoRa, SIGFOX, and even 5G, which will be the basis of long-range IoT connectivity. Batteries: If the sensors cannot be connected to fixed power sources, devices powered by batteries have to be used, as these have a limited charging capacity. First, assess the passive or active sensor characteristic based on the necessary capture frequency and the required battery capacity. Processor: Assess the criteria of data processing capacity, flash memory for storage, the space it occupies (reduced sizes prevail), low energy consumption for data communication, and cost. Management platforms: Platforms have been developed to connect sensors relatively easily if the company does not want to develop its platform. Cost of the sensors: It has been reduced in recent years, despite some particular ones being still expensive. Characteristics of the sensors: Taking into account the precision and speed required for the measurement, we need to perform and the optimal obtaining frequency. Then, the search for the most suitable sensor is carried out.

Intrinsic error in the configuration of the sensor system. It is important to consider the optimal installation characteristics (place, barriers that can generate interference, type of support, and others) to achieve the proper functioning of the sensors. · Errors associated with the sensor (reading, precision, communication, interpretation): Loss of information because the reading is incorrect, and the margins of error are higher than required and can lead to misinterpretation of the data and take actions that are not appropriate. · Incompatibility of systems: Establish a single system or compatible systems when collecting all the required information. If it is not foreseen from the beginning, situations can arise in which certain sensors cannot be connected, or intermediate devices must be used to achieve this compatibility. Computer security in IoT. Beyond what is mentioned in the General Guide, IoT systems are physical sensors connected to platforms where a large amount of sensitive data is processed and can be a vulnerable entry and exit point. For this reason, they must have rigorous security protocols, in which security prevails over cost [5].

13.7.2 Challenges and Challenges in Organizations

News of any kind, and also technological ones, opens the door to numerous advantages and uncertainties about how they can affect us in other aspects. Below is a series of reflections on different areas still to be explored related to IoT technology. The IoT is an opportunity for companies to improve their ORP performance through new practices based on continuously measuring and "monitoring" many useful parameters for ORP. Proper management of information according to the sequence "measure-record-analyze-act-review" allows both to generate immediate action processes (identification of dangerous situations, issuing warnings and alerts, activation or inhibition of equipment or processes, etc.), such as collectively analyzing a large volume of data to help decision-making, including predictive prevention. Therefore, for the implementation of the IoT, the company needs to have a global vision when establishing the objectives to be achieved, identifying the aspects to be supervised/controlled and the necessary measurements for this purpose, selecting the most appropriate technology, and defining the derivative actions and their criteria, including the adaptation of procedures and work processes.

Involvement in the Forms of Work: The IoT in occupational risk prevention seeks to support workers in preventive aspects, safeguarding their safety at all times without perceiving unnecessary interference in their work activity. It should be noted that, despite the fact that IoT systems can provide additional protection, their use must be framed within the principles of preventive activity and attend to their priority (elimination or reduction of risk, collective protection, use of PPE, safety procedures). On the other hand, while this technology will mainly deal with monitoring control tasks, those responsible for prevention will add value to companies by defining processes and managing security actions. For this reason, they must change their approach and spend less time on continuous surveillance and control of potential risks and more on managing action protocols when the systems generate alarms. Responsibility and compromise. The real-time recording of information and/or warnings of certain situations (use or misuse of equipment or PPE, parameters exceeded in areas or machines, unauthorized access, inappropriate behavior, etc.) increases the company's ability to know the reality about the presence of hazards or performance deviations. This must imply greater responsibility and commitment on the company when establishing actions and defining protocols and responsibilities for action that allow redirecting the risk situations that are detected.

Privacy: Since the sensor system can continuously take a great variety and quantity of personal data, its correct treatment is especially relevant. It is assumed that the data will always guarantee workers' safety and accident prevention. Therefore, its use and, in particular, the purpose of its collection must be perfectly delimited and defined, establishing protocols that guarantee data protection staying under the legal criteria established by the regulations. In this way, said sensitive personal data is protected, and the staff knows the purpose of its treatment. Managed this part properly, IoT systems can contribute their beneficial part to people in their protection.

Specific risk assessments for new devices need constant measurement. Its associated technologies introduce new risks; therefore, the quality and safety of the incorporated products must be ensured by carrying out specific risk assessments that guarantee the control of potential risks. Just as it would not be appropriate to use any helmet to protect ourselves from certain impacts, it would not be appropriate to use any smart-helmet to protect us from danger. Therefore, it is advisable to conduct specific evaluations that analyze the specific risks and design preventive and corrective measures. Predict to prevent is what an IoT system combined with artificial intelligence will record a large volume of data from various sources and put them together like pieces of a puzzle to conclude and thus predict events. From the reading of the circumstances of the people, of the company (of the environment, security data, etc.), together with those of other sources, it can establish probabilities of events and predict and identify that certain conditions and or behaviors may be likely to cause a type of accident. These data can be incorporated into risk assessment methodologies, providing objective information for assessing the frequency and probability of a risk appearing.

13.7.3 IoT Security

IoT objects are always at risk since there are no impregnable systems. However, with the right tools, this risk can be greatly reduced [6]. Here are some of the basic tools to improve IoT security:

- Acquire reliable hardware and software. Acquiring the most affordable can put us in a vulnerable place. Consider that IoT objects such as an activity bracelet or a company computer collect a lot of potentially dangerous information if it spreads.
- Buy only updated and upgradeable devices. Being able to fix security issues is key. That is why, it is imperative to only invest in devices that can patch undiscovered vulnerabilities as they become public.
- Devices that allow changing the password. Security completely leaves out IoT objects without a password and those that do not allow changing it. Therefore, we must always look for systems to update it frequently.
- **AntiDDoS Systems:** There are protection systems on the market that prevent IoT devices from being vulnerable to DDoS attacks, turning into botnets or zombie networks.

13.8 FUTURE OF IoT

The percentage of companies that include IoT solutions in their processes grows year after year, but not only that, but the number of devices connected to these solutions also expands. As a result, IoT technology is expected to become a potentially growing business in the coming years. 79% of those who have already adopted IoT solutions predict that 50% of business processes will include IoT sensing. Derived from this, the volume of data handled will be large enough (big data) to include the use of artificial intelligence (AI) and machine learning to analyze all this data and generate intelligent actions [7].

This technological breakthrough will not be overlooked by most large companies that, according to 82% of IoT users, will seek to establish collaborations with industries to develop IoT solutions that improve their businesses and the capacity of analysis in their processes. Certainly, the Internet of Things is the future. In a few years, this is likely to be referred to as an inconspicuous concept, if not an everyday part of how individuals and businesses use technology and the Internet, just as searching for something on the Internet is today. However, perhaps one of the most exciting things about IoT is that it is just getting started, and every day, individuals and businesses are finding new uses for this technology [8].

13.9 CONCLUSIONS

This new scenario presented to us with the Internet of Things emphasizes that the world is advancing faster than is believed. In our times, the dynamics are much more intense and complex; in the history of humanity, never before have the present and the future been so close to each other, almost overlapping on the timeline. That is why, especially in health, imagining the future is to talk about the present and understand the new digital dimension that is already a reality. We are clear that the reflections on this topic point to the great opportunities with massive data and the challenges around security and privacy, complex issues that transcend the field of health and require interdisciplinary approaches. IoT represents the area of health, epidemiology, and public health, innovative possibilities to understand many phenomena in the health area thanks to algorithms and artificial intelligence. However, one of the main challenges will be accumulating data to transform them into research and action in health in real time.

So, the Internet of Things can be a digital space where the health area gravitates in the same orbit with technological advances. However, at the same time, hand in hand with public policies and appropriate regulatory mechanisms benefits society by optimizing individual health and public health.

With the review of the state of the art of IoT and artificial vision, it has been shown that studies and technological advances in the not too distant future will be implemented in people's daily lives, and they will be able to live with it. These technologies seek to provide comfort and facilities in the different environments in which we live.

Research is already being criticized for its wide range of adoption; however, it is improbable to address the challenges in its development and provide privacy and security confidentiality to the user. Be a ubiquitous technology. IoT deployment requires strenuous efforts to address and present security and privacy threats solutions.

REFERENCES

1. Meyer, M. (2021, May 6). Deep Learning for Internet of Things. IEEE Access. https://ieeeaccess.ieee.org/closed-special-sections/deep-learning-for-internet-of-things

2. Newman, P. (2019, January 28). IoT Report: How Internet of Things technology growth is reaching mainstream companies and consumers. Retrieved from https://www.iwec-foundation.org/news/internet-of-things-reaching-mainstream-companies/
3. Gatsis, K., & Pappas, G. J. (2017). Wireless control for the IoT: Power spectrum and security challenges. In *Proc. 2017 IEEE/ACM second International Conference on Internet-of-Things Design and Implementation (IoTDI), Pittsburg, PA, USA, 18–21 April 2017.* INSPEC Accession Number: 16964293.
4. Sfar, A. R., Zied, C., & Challal, Y. (2017). A systematic and cognitive vision for IoT security: A case study of military live simulation and security challenges. In *Proceedings of the 2017 International Conference on Smart, Monitored and Controlled Cities (SM2C)*, Sfax, Tunisia, 17–19 Feb. 2017. https://doi.org/10.1109/sm2c.2017.8071828
5. Kumar, J. S., & Patel, D. R. (2014). A survey on internet of things: Security and privacy issues. *International Journal of Computer Applications*, 90(11), 20–26. http://faratarjome.ir/u/media/shopping_files/store-EN-1520245899-5958.pdf
6. Soumyalatha, S.G.H. (2016, May). Study of IoT: understanding IoT architecture, applications, issues and challenges. In *1st International Conference on Innovations in Computing & Net-working (ICICN16), CSE, RRCE. International Journal of Advanced Networking & Applications* (No. 478), Bengaluru, Karnataka.
7. Agrawal, S., & Vieira, D. (2013). A survey on internet of things. *Abakós*, 1(2), 78–95.
8. Macaulay, J., Buckalew, L., & Chung, G. (2015). *Internet of Things in Logistics: A Collaborative Report by DHL and Cisco on Implications and Use Cases for the Logistics Industry* (pp. 439–449), DHL Trend Research and Cisco Consulting Services.

14 Next-Generation IoT and the World of Sensors

ABBREVIATIONS

AI Artificial Intelligence
NGIoT Next-Generation Internet of Things

14.1 INTRODUCTION

The Internet of Things (IoT), along with artificial intelligence (AI) and Big Data, is at the center of the digitization of the world economy; the data obtained from sensors can be monitored and fed back to act, save the information, or communicate with another connected object hundreds of miles away. So it is possible to avoid unnecessary communication and storage costs while using machine learning and AI to identify data patterns that impact physical or business processes. The Next-Generation Internet of Things (NGIoT) focused on Cloud to Edge to IoT emphasized the need for an open industrial platform for the coordination of the elements at the edge of the cloud, and focused on the technological challenges and competitive impact of European stakeholders in its relation to the data economy, needs to define a strategy for the future IoT and edge computing, with a market projection of more than 5 years. The consensus was reached in general about the need for reliable IoT and edge computing platforms and integration mechanisms to support this next stage of digitization. The challenges and opportunities ahead require establishing a shared strategic vision for NGIoT and edge computing in the world. The NGIoT uses intelligent solutions integrated into the edge based on high connectivity, processing capabilities for edge devices, and analysis of information in real time, which is possible by some convergences of technologies as key information and communications technologies , such as hyperconnectivity, new network architectures, and edge computing. According to Cisco, by the year 2030, there will be about 500 billion devices, and it is estimated that by 2030, 90% of vehicles connected are transforming the world into an intelligent one where users can quickly access information or obtain data from our environment in a faster way [1].

This chapter explains recent trends in NGIoT technologies and finds out more about the modern sensors driving this technology as building blocks for new generations of IoT devices. At the beginning of the era, technologies were limited to specific sensors since, despite their great utility and their Internet connectivity between devices, the information provided from the environment did not have intelligent capabilities or information processing intelligent results. However, the advancement in IoT technology has placed it as a hub in connectivity technology that will reach all parts of the world.

DOI: 10.1201/9781003302902-14

The indisputable and preferable use of IoT technologies [2] has been promoted at great speed and great while other technological areas, such as AI, among other concepts, advance due to its multiple advantages. Therefore, this chapter will provide us with more information on understanding the underlying concepts and technologies to further our understanding of NGIoT and its relationship to sensors and advancements, from the basic use of IoT to NGIoT and its relationship with the evolution of sensors.

14.2 THE WORLD OF SENSORS

Passing from the physical world to the digital world is a process realized by integrating fog computing and IoT-based circuits, providing the environment required for building a new ecosystem of services where sensors are fundamental.

Sensor networks are also increasingly vulnerable to attacks that occur in IoT infrastructures. As Pacheco and Harin explain, "However, and given that fog computing directs services from the edge of the network, its integration into the IoT infrastructure, in addition to optimizing latency and quality of services, has mechanisms that allow the integration of security measures".

So, redundant test systems are robust, effective, and predictive against cyberattacks regarding reactive systems that act when attacks are realized. However, these experts say [3], "in the ecosystem of the Internet of things, the security and anomalous behavior of sensors and other IoT components have to be determined by more complex orders until there are no security components that protect the system from own sensors that communicate the real world and the digital one". And virtually every device contains with some degree of "intelligence". Most are based on autonomous, programmable software. "In an ideal world, the security that is built into devices and sensors would have to make them more resilient. However, the reliability and security of IoT applications, sensors, and elements can only be improved to a certain extent before the deployment of the system. The reasons vary and can come from restrictions related to electricity consumption, the complexity of the infrastructure itself, an inadequate IoT design, or a lack of planning".

14.2.1 Great Growth Potential

Sensors have the potential to operate autonomously. Janus Bryzek, the "father of sensors", says, "Systems can learn and change their future behavior, giving rise to the creation of more intelligent devices and programs. In parallel, the combination of computational power, advanced algorithms, and data sets that arrive massively to feed said algorithms are giving rise to the birth of a new era". And the acceleration of the mass adoption of the IoT ecosystem [4] is due to different causes: "First of all, because the new Internet protocol IPv6 is going to allow a practically unlimited number of networked devices and sensors". The expert adds as another decisive reason the fact that large IT providers offer IoT support, "applying modifications to their networks, adding fog technology, and probably also Swarm technology, thus reducing the complexity and lower costs of network connectivity", taking into account GE estimates, which speak of an industrial Internet with a potential to add 15 quintillion dollars to the global gross domestic product in the next 20 years (and Cisco that raises to 19 quintillion dollars), your prediction of the value of IoT is the most significant growth ever in human history.

14.3 FUNCTIONS OF THE SENSORS THAT INCORPORATE THE OBJECTS WITH IoT TECHNOLOGY

A sensor is a compound capable of detecting, sizing, or indicating variations in the environment into a specific physical quantity, either temperature, humidity, or pressure, and converting them into electrical signals. Signal conditioners amplify this modification and convert it into a readable format, as shown in Figure 14.1. They are usually part of a system, the electronic circuit contains various components, such as a controller, a source of power, and a memory module, is interconnected to collect, process, and send information over the network [5].

Figure 14.2 shows different types of sensors, grouping them according to their functionality.

14.3.1 Sensors for Temperature

These sensors allow knowing "the amount from Energy thermal that allows detecting a physical change in temperature of a particular source and convert the data to a device or user" [6]. Temperature sensors have been widely used in air-conditioning

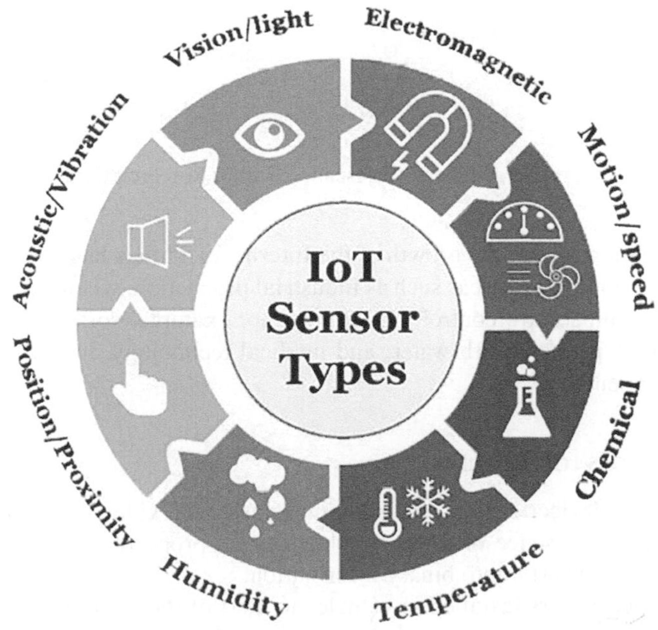

FIGURE 14.1 IoT sensor types.

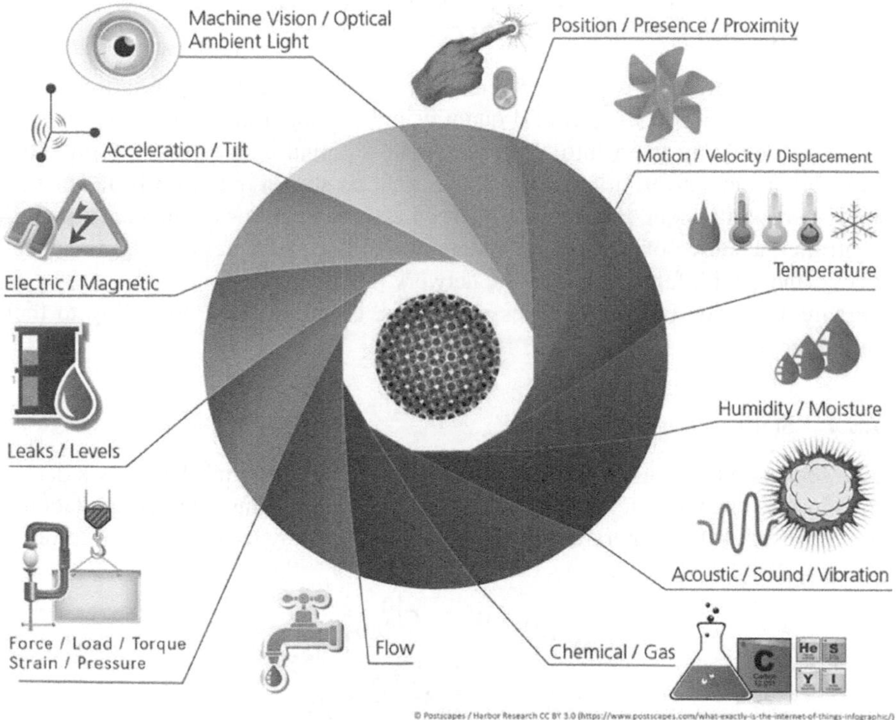

FIGURE 14.2 Sensors and actuators. (Postscapes/Harbor Research CC BY 3.0.)

systems or appliances, but the growth of the Internet of Things has opened new possibilities for monitoring in areas such as industrial production, where some machines require strict temperature control; agriculture, for example, soil temperature conditions and how plants absorb water; and medical technology, including telemetry systems for patients.

14.3.2 Sensors for Proximity

The use of these devices is very important in security and video surveillance. These sensors sense the presence of a nearby object or its properties, and if they detect something, they convert it into binary information.

For example, sensors installed in vehicles to alert of the presence of objects, the devices installed in smart parking or devices that help improve the shopper experience by offering a discount on some products that they are interested in.

14.3.3 Sensors for Measurement Gases

They are increasingly used in urban environments to provide information on the level of contamination; that is, they measure the quality of the air, and thus collect information on air pollution experienced by residents. Therefore, they must be well-calibrated, going through a laboratory calibration process.

14.3.4 Sensors for Pressure

They are used in industrial installations, maintenance of water systems, or obtaining weather forecasts. These sensors monitor systems and equipment of pressure. Upon deviation from the standard range, the device notifies the system administrator that there is a problem.

14.3.5 Sensor for Moisture

These sensors are important in agriculture. For example, these sensors allow to improve the efficiency of irrigation systems and improve the management of natural resources

14.3.6 Sensor of Level

These sensors take information from various points in a container and perform the measurement, taking into account the top edge and the bottom of the container [7].

Among other smart sensors seen today, it is possible to name some marketed and used in certain population sectors and industries.

- Remote level monitoring sensor
- Parking space sensor
- Global positioning system sensor vehicle recovery and asset control
- Sound detection sensor and sound pressure level measurement
- Community safety solutions
- Video surveillance solution in the cloud
- Vibration sensor
- Hydrogen sulfide sensor
- Smart water meters
- Body temperature sensor
- Temperature and humidity monitoring
- Detection and location of assets and movements
- Human motion detection
- Home security
- Button to order service, products, or assistance

14.3.7 Intelligent and Autonomous Sensors

An intelligent sensor is a device that allows monitoring and sending data of different variables at a preestablished frequency to a storage and analysis platform. Different data transmission technologies: Wi-Fi, 4G, a low-power wide-area network, and

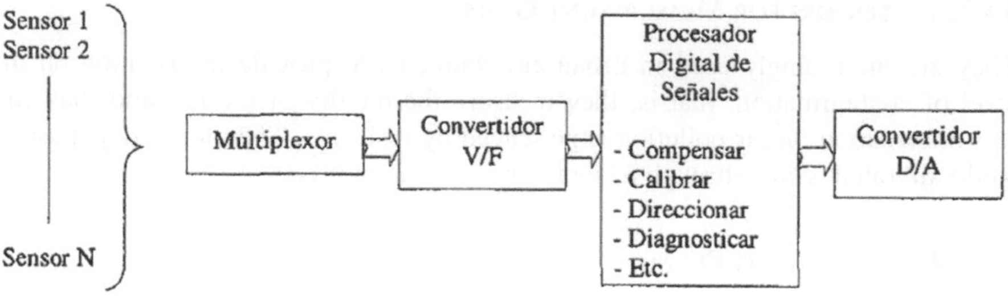

FIGURE 14.3 1980s generation smart sensor [5].

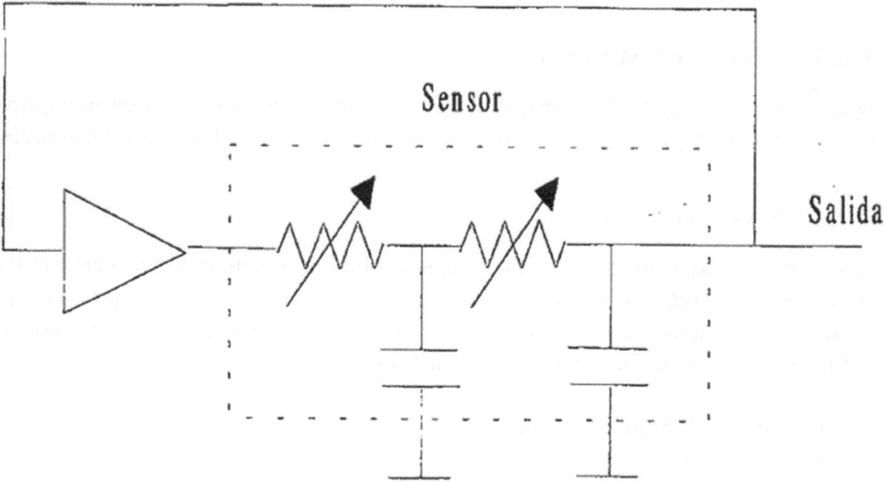

FIGURE 14.4 The first smart sensor solves the temperature compensation problem [5].

general packet radio service, among others, can be used. Sensors are the basis through which relevant information, trends or analytics, and reports are obtained that allow better decision-making and, in general, improve profitability [8].

In the beginning, smart sensors have their starting point, Figure 14.3, with the sensor that solves the problem of temperature compensation in the elements being fed back and connected to an inverter to create an oscillator, demonstrating its operating structure through the circuit of Figure 14.4.

Smart sensors have been an exciting topic since their inception and have been improving in recent years; now, we show some of the essential details, such as the features and uses of smart sensors. For example, smart sensors feature secure data transmission, no malware, no viruses, automatic device update, and data migration and integration, making it possible to control IoT ecosystems with smart sensors. Modern sensors are driving the advancement of NGIoT; as mentioned earlier in this document, a smart sensor is such a device that allows monitoring and sending telemetry data to remote servers and help with smart decision making based on the logics running on remote servers.

14.4 THE NEXT-GENERATION INTERNET OF THINGS

If we consider the exponential growth of connected devices and systems, processing and analyzing data are becoming the driving forces behind the digitization of the economy, society, and environment. As devices become more and more "Smart" to collect data from transmitting information Y Activate Actions in real time, the IoT is at the center of this digital transformation, integrating devices, data, computing power, and connectivity [9]. The IoT creates environments smart with digital technologies to optimize the shape in which we live our lives. In the near future, more than 41 billion IoT devices are expected to be launched (International Data Corporation). This will lead to exponential data growth, pushing computing operations and data analysis to the limit [10] (Figure 14.5).

"Edge computing" is based on a multilayer technology that allows the management and automates devices IoT connected. Is the evolution logic of a model dominant among cloud computing features?

> Filters the transfer of essential data from the system to the cloud, Supports resiliency, real-time operations, and security, privacy, and protection. At the same time, reduce consumption from Energy. In edge computing, processing moves from one point centralized, plus near (or even within) of own IoT device: the "Edge" or the periphery of a network [11].

14.5 OPPORTUNITIES, CHALLENGES, AND SOLUTIONS

By 2030, the forecast is that each person will have fifteen devices connected, so IoT is transforming the world into a place where it will be possible to access everything quickly. In general, industries will engage in quick speculation to deal with companies according to the development level of IoT technologies. The IoT is expanding radically. Several researchers have constantly been developing new strategies to adapt to these

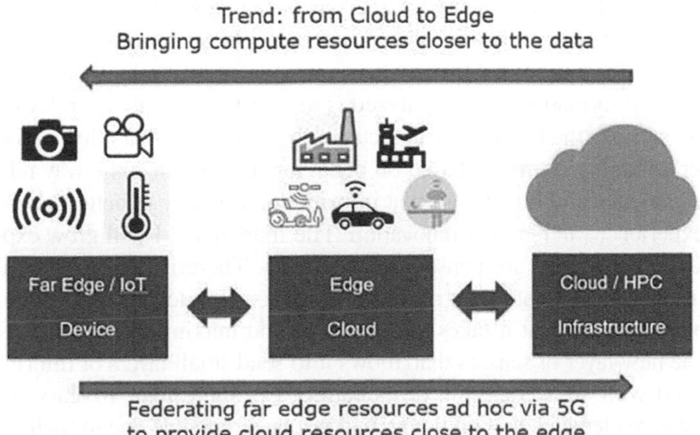

FIGURE 14.5 IoT Data Flow NGIoT.

phenomena. IoT applications are incredibly diverse and incorporate multimedia in weather real, systems from health based on IoT, industries' IoT-based next-generation smart devices, and next-generation smart agriculture IoT-based generation. In addition to security requirements, meeting the needs from the applications is a homework review; living in a situation epidemic requires a quick analysis of data and mechanisms from prediction. The problems and challenges related to access technologies, such as spectrum scarcity, are another critical concern regarding sharing optimal resources between many IoT devices. Using AI-based solutions to enable dynamic and adaptive technologies is one of the possible ways to advance in this area [12].

Consequently, this special issue has attracted researchers from academia and industry to investigate the opportunities of IoT-based user scenarios of NGIoT and study its effect on solutions to problems and challenges discussed previously and propose solutions viably. As a result, there is a contribution from researchers in different areas of interest related to the applications based on IoT from NGIoT that includes the following:

- NGIoT-based smart agriculture
- NGIoT-based smart cities
- NGIoT-based smart health
- NGIoT-based industrial IoT
- NGIoT-based data analysis
- NGIoT-based spectrum sharing techniques last generation
- NGIoT-based media

IoT is the technology that profoundly changes the way in which humans relate to machines and machines to each other, and it is going to be the real engine of digital transformation today. They are mobilizing almost all industries to take advantage of existing technology to digitize your processes and connect your production elements, which consist of various elements as follows: the first element is the connection to the network of those sensors and those machines which is a communication block; the second is a cloud storage block because all those machines start to send massive amounts of information to be able to be analyzed; the third has that capacity for analysis or big data; and the fourth is a security component because effectively this information has to be secure. Extreme big data is an essential enabler for this so that the machines send information by themselves. It is not enough that the information that is analyzed is to select the one that is relevant to take a decision and activate the technology that currently exists but what allows it is to predict with greater reliability estimates based on behaviors that are in some way related to the customer experience and transfer all that information capacity to better income, greater customer experience, and greater innovation. The market and I will grow exponentially because new technologies are constantly appearing. There is one in particular that is very relevant called at night and it is based on the existing technology of the operators from the point of view that it takes advantage of and maximizes its coverage; second, introduce the new layer of sensors that allows it to send small layers of information very efficiently and with an average life of its batteries of more than 10 years. This allows you to sensualize elements that until now had not been possible due to their volume and capillarity. Electricity meters and water meters will make the distribution much more efficient. Energy distribution is the technology that makes it accessible for everything that contributes a differential value to society or the economy to be connected, and that is what is going to lead us to multipliers of millions of connected elements in the coming years in the world. Our communication networks have always distinguished themselves by providing the greatest reliability in data, and they are one of the great actors in the cloud trio. We have great analytical capacity as a result of many years of experience working with end-customer data, and lastly, we are also one of the few players, I would say practically the only one, that is capable of providing extreme, extreme security on all the blocks. We are helping many companies to transform themselves to start business models to have the closest action with their clients to be more efficient, and what we are also guaranteeing is that these companies will have a better future because this technology has only just started and is going to be the real engine of digital transformation.

14.6 CONCLUSIONS

IoT has become the most valuable and popular technology, developed faster under quality of experience requirements and quality of service of users adopting it daily. Therefore, academia and industry are improving and developing new procedures to adapt to these phenomena. On the other hand, NGIoT applications are incredibly different, incorporating smart homes, smart health care, smart industries, real-time multimedia, and smart agriculture; addressing needs, satisfaction, and safety issues with AI solutions to make technologies robust and adaptable to important tasks. The potential uses of NGIoT are increasing, the effect of IoT developments from the next generation in solutions to the challenges mentioned above propose viable solutions.

This chapter presents information that covers topics of interest that will impact the future of data analytics, smart cities, smart agriculture, health care based on IoT; real-time multimedia applications based on IoT; and smart industries; are working to exchange from spectrum based on IoT to NGIoT – supported with techniques from security and privacy [13].

The market of smart IoT sensors studies various aspects of global and regional industries in size, share, product status, latest business trends, and forecast. Also, it provides detailed information on competitors and specific growth opportunities with crucial IoT smart sensors industry drivers, IoT smart sensors market progress, and related approaches to IoT smart sensors market considers aspects of industry and segments, where the global IoT smart sensors market size is estimated to grow by 22% between 2022 and 2030. Furthermore, global IoT smart sensors market segmentation considers company profiles, region (country), major type, application, stakeholders, and other industry participants.

REFERENCES

1. Kim, G., & Jung, I. Y. (2019). Integrity assurance of OTA software update in smart vehicles. *International Journal on Smart Sensing and Intelligent Systems*, 12(1), 1–8. doi: 10.21307/ijssis-2019-011.
2. Gillis, A.S. (2021, Aug 13). What is IoT (Internet of Things) and How Does it Work? IoT Agenda; Tech Target. https://internetofthingsagenda.techtarget.com/definition/Internet-of-Things-IoT

3. Kalsoom, T., Ur-Rehman, M., Ramzan, N., & Ahmed, S.(2020). Advances in sensor technologies in the era of smart factory and industry 4.0. *Sensors*, 20(23), 6783. doi: 10.3390/s20236783.
4. Mukhopadhyay, S. C., Tyagi, S. K. S., Suryadevara, N. K., Piuri, V., Scotti, F., & Zeadally, S. (2021). Artificial intelligence-based sensors for next generation IoT applications: A review", *IEEE Sensors Journal*, 21(22), 24920–24932.
5. Shafique, K., Khawaja, B. A., Sabir, F., Qazi, S., & Mustaqim, M. (2020). Internet of things (IoT) for next-generation smart systems: A review of current challenges, future trends and prospects for emerging 5G-IoT scenarios. *IEEE Access*, 8, 23022–23040.
6. Gorai, S. (2018). Smart sensors slideshare.net Retrieved on January 31, 2019, from https://es.slideshare.net/SupriyaGorai1/smart-sensors-94653798/13
7. Villalba-Diez, J., Schmidt, D., Gevers, R., Ordieres-Meré, J., Buchwitz, M., & Wellbrock, W. (2019). Deep learning for industrial computer vision quality control in the printing industry 4.0. *Sensors*, 19(18), 3987.
8. Veres, M., & Moussa, M. (2020). Deep learning for intelligent transportation systems: A survey of emerging trends. *IEEE Transactions on Intelligent Transportation Systems*, 21(8), 3152–3168.
9. Liang, F., Yu, W., Liu, X., Griffith, D., & Golmie, N. (2020). Toward edge-based deep learning in industrial Internet of Things. *IEEE Internet of Things Journal*, 7(5), 4329–4341.
10. Farivar, F., Haghighi, M. S., Jolfaei, A., & Alazab, M. (2019). Artificial intelligence for detection, estimation, and compensation of malicious attacks in nonlinear cyber-physical systems and industrial IoT. *IEEE Transactions on Industrial Informatics*, 16(-4), 2716–2725.
11. Ma, Y., Jin, J., Huang, Q., & Dan, F. F. (2018). Data preprocessing of agricultural IoT based on time series analysis. In *Intelligent Computing Theories and Application (Lecture Notes in Computer Science)* (Vol. 10954), Cham: Springer. doi: 10.1007/978-3-9-95930-6_21
12. Saravanan, M., Kumar, P. S., & Sharma, A. (2019, July). IoT enabled indoor autonomous mobile robot using CNN and Q-learning. In *2019 IEEE International Conference on Industry 4.0, Artificial Intelligence, and Communications Technology (IAICT)* (pp. 7–13), Bali, Indonesia.
13. Kirwan, C., & Zhiyong, F. (2020) *Smart Cities and Artificial Intelligence*, Amsterdam: Elsevier.

15 Artificial Intelligence and Networking

ABBREVIATIONS

AI	Artificial intelligence
ANN	Artificial neural networks
B5G	Beyond 5th Generation
IBN	Intent-based networking
ML	Machine learning
QoS	Quality of service
RL	Reinforcement learning
SDN	Software-defined networking
SOM	Self-organized maps

15.1 INTRODUCTION

Artificial Intelligence (AI) is the future; it is an emerging and very powerful technology that has a significant impact in numerous fields. With the rapid development in science and technology, it directly relates to upgrading those relevant domains touching our lifestyle with giving a business goal or commercial purpose to the legacy technology and upgrades to next-gen technology. Computer networks are also one very critical part of our lifestyle that is touched by technological developments and keeps itself growing day by day. AI, being the technology which is seemingly a great cash cow, has created its own space and is undoubtedly a new touchpoint to the boundaries of computer networks. Many organizations are expending a lot for leveraging AI to simplify the complex jobs in the day-to-day activities in their respective domains. At the same time, the use of AI in networking technology has exerted a profound influence in many other areas. It is worth mentioning that AI has a very significant impact on futuristic networking [1]. The conjunction of AI with computer networks will be a good composition to offer a predictive, dynamic, self-diagnostic, adaptable, and auto-healing network. It is something that will boost the automation in networking and supposedly less dependence on manual methods for control, operations, and maintenance of computer networks [2].

At the same time, with the evolution of cloud computing and the software-defined networking, AI is fueling the current interest in network programmability for improved network automation in the dynamic and agile infrastructures, including edge and fog environments. At one end, designing of network considering AI and ML techniques helps in optimization, automation, control, and management of network. Also, it addresses the complementary topic of supporting AI- and ML-based systems through novel networking techniques, including new architectures and algorithms

DOI: 10.1201/9781003302902-15

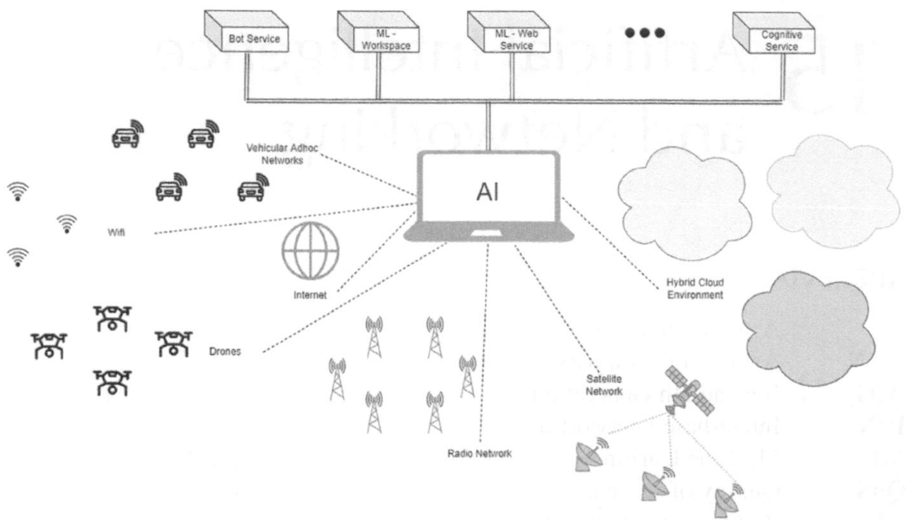

FIGURE 15.1 AI in networking.

to optimize network support for architecture and design of data communication and processing for AI purposes [3].

The COVID-19 pandemic has accelerated the urge for automation and digitalization of industries and has shown the effectiveness of AI in this direction. The world has started adopting a new business model with certain accelerating trends where businesses will be more virtual, business interactions will be more online, and a larger number of employees will prefer working remotely. Telecom Service Providers and the network operators are the backbone of this digital transformation and the Internet-based business model. The pressure due to the pandemic has increased on the telcos with the demand for data and voice communication increasing and the need to extend the network capacity rapidly [4]. They were also supposed to support the government and the health-care systems by providing high-speed connectivity. AI is being considered as the central point or the core of the telcos transformation into this new reality of the world post the COVID-19 crisis, because the AI will help in delivering of the superior performance in short and long term in a flexible and agile environment (Figure 15.1).

15.2 WHAT IS AI?

AI is something which can have its place everywhere nowadays, such as from the gaming stations to maintaining complex information at work. Computer engineers and scientists have been working for more than two decades in this area of making intelligent systems to impart intelligent behavior on the machines making them think and respond to real-time situations. AI has been so popular among computer scientists and has been defined differently among the fraternity but in simple terms it can be understood as that department of science and technology which makes intelligent computer systems that can simulate the human intelligence processes by machines.

It is usually considered the science of making computers perform tasks that require humanlike intelligence, combining computer science and robust datasets to enable problem-solving. We are referring the definition given by Stanford Researcher, John McCarthy [5], "Artificial intelligence is the science and engineering of making intelligent machines, especially intelligent computer programs. Artificial Intelligence is related to the similar task of using computers to understand human intelligence, but AI does not have to confine itself to methods that are biologically observable". AI in general is a broader concept of machines being able to carry out tasks smartly has been categorized into subfields of machine learning (ML) and deep learning, which are very popular and frequently referred in conjunction with artificial intelligence. AI and ML are those segments of computer science which are correlated with each other in many ways and are the most trending technologies that can be used for creating intelligent systems. Although these related technologies are sometimes used as synonyms for each other, both are having its specific place and in general ML is considered as a subset of AI.

These disciplines of machine learning and deep learning are comprised of several AI algorithms required for the creation of expert systems to make predictions or classifications based on input data. ML is probably the most popular application of AI enabling machines to learn from large amounts of data and act accordingly without the need of programming machines explicitly. However, deep learning can be considered as a special type of ML that studies artificial neural networks (ANNs) containing more than one hidden layer to "simulate" the human brain [6]. Deep learning is one of the most widespread ML methods at this time having successfully applied to different fields such as the computer vision, speech recognition, and bioinformatics.

15.2.1 Machine Learning

Machine Learning (ML) is an application of AI, and we can think of it as a science of enabling machines to learn things from past cases. Learning happens from the past data or historical records to produce reliable results, and the data used for machine learning is generally referred as the training data. ML is done based on some method or algorithm that involves the process of using some mathematical model of data to help a computer learn without any direct machine instruction or explicitly programming [7]. Based on that learning, machines can predict or take decisions such that if some behavior existed in the past (in the training data), and then machines will be able to predict if it can happen again [6]. The overall process of learning is such that it enables a computer system to learn by itself and to keep improving the model on its own, based on the machine learning experience. Generally, computer science and statistics are used in ML algorithms for the prediction of rational outputs. Complex algorithms are used for machine learning that constantly keep iterating over large datasets to analyze the patterns in data, enabling machines to respond. Based on that learning, machines may respond to different situations for which they have not been explicitly programmed. ML may be applied to several of the critical issues or tough decision-making things that involve a large number of datasets for analysis and making predictions based on that historical data, such as credit card fraud detection,

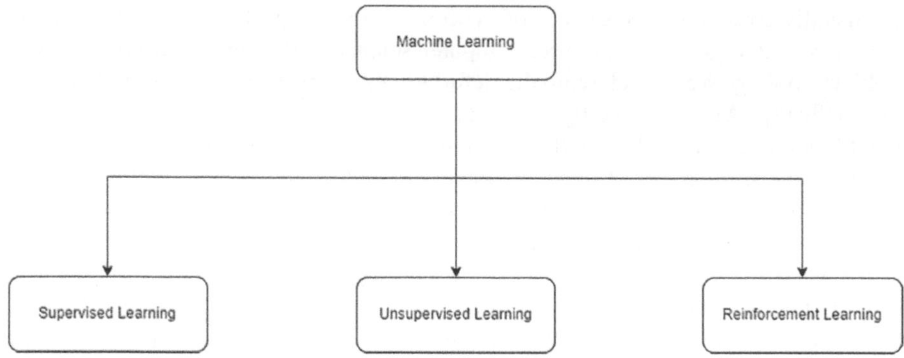

FIGURE 15.2 Types of ML.

loan prediction for a bank customer, enabling self-driving cars and face detection and recognition. As per mentioned in Figure 15.2, there are three major types of ML.

Supervised Learning: It is a ML type that has the learning model getting trained based on a labeled dataset, where labeled dataset is the dataset having both input and output parameters. The labeled datasets are provided to the system to get the model trained, and then the model will produce an inferred function to create mapping for new examples. Examples of credit card fraud detection are good examples of supervised learning algorithm.

Unsupervised Learning: It is such ML type that uses unlabeled data for training the machine learning model. The algorithms are much complex because of the fact that the data to be fed is neither classified nor labeled and the machine has to learn on its own without any manual supervision. The algorithm has to group unsorted information based on similarities, patterns, and differences without any prior training of data or without providing any correct solution of any problem. An example like recommendation engines for certain social media applications or e-commerce sites for the suggestion mechanism is a good example of unsupervised learning.

Reinforcement Learning: It is a type of ML that has software agents and machines that help in determining automatically the ideal behavior within a specific context, to maximize its performance. The algorithm improves itself and learns from new situations using a trial-and-error method, where the algorithm has to encourage or reinforce the favorable outputs, and to discourage or punish the nonfavorable outputs.

15.2.2 Neural Network

A neural network or artificial neural network (ANN) is a series of algorithms that explain the behavior of the human brain and mimic its processing with interconnected units called as neurons. ANN has become a popular framework to perform machine learning methods that use learning algorithms inspired by the brain to store information. This helps in recognizing patterns and solving complex problems in the fields of AI, machine learning, and deep learning. ANN algorithms are the mathematical models that process information by responding to external inputs, relaying information between each unit that requires multiple passes at the data to find connections and derive meaning from undefined data. A "neuron" is a mathematical function that is used to collect and classify information based on a specific architecture. The architecture representing connections with other neurons is the neural network that strongly resemblance to statistical methods such as curve fitting and regression analysis.

15.2.3 Deep Learning

The technological concepts of AI, ML, ANN, and deep learning are very much interrelated to each other with no clear-cut demarcations between them. We can think of deep learning as one popular technique of a broader area of machine learning. Deep Learning is that machine learning technique which takes inspiration from the functioning of the human brain where the learning models are built on top of ANN. The term "deep" in deep learning refers to the learning technique having multiple layers in the neural network. It uses huge neural networks that have many layers of processing units, with machines having advanced computing power and improved training techniques to learn complex patterns in large amounts of data. It is used for solving complex problems with having data that is huge, diverse, and less structured. Usages such as image recognitions, speech recognitions, automated driving, industrial automations, medical research, aerospace, and defense have common applications of deep learning.

15.3 HOW AI CAN TRANSFORM COMPUTER NETWORKS?

Communication is an essential and prime need of human beings, and from its many means, digital communication has become one very important method at this stage of world. To fulfill this, the computer scientists, researchers, and the communication industries are always searching for new and latest network technologies which can be effortless, noninterrupting, efficient, stable, automatic, and seamless connectivity. The future networks are supposed to be always-connected networks with auto fault-isolation and resolution feature, and the network devices will convey network health to each other to make stable connections. There is also a need of universality among the different types of networks to make all connections possible where one should not have to worry much whether using a Wi-Fi network, a mobile network, Bluetooth, or one of the many IoT network technologies and will just work.

Considering AI and its behavior into networking, we can think of a universal network that can make decisions for the user based on user's application demand, current location, and activity, to make seamless connections with the best-possible network to handle the network traffic without interrupting user's experience. With AI, the network can dynamically take on-demand routing decisions. The expectation with AI is that it makes all of this possible and will be the facilitator of real-time conversations between networks. At the same time, with the help of AI, it can be ensured that all network traffic will receive the best-possible quality of service out of the connections available to them. AI will supposedly facilitate in fulfilling these networks to exist across all industries and borders.

With incorporating the AI/ML methods and algorithms in designing a network, it will create a new approach to design of network, but it will still need a human support to create such a network and the human factor will not go away. Where AI may be able to give new methods for designing cost-effective and faster networks, however, the actual network building must still be carried out by humans. AI will give an edge or a boost to human intellect and creativity with its massive computing power to create new design and management techniques that humans could not build on their own. AL and ML algorithms of having self-improving capabilities can also be leveraged to improve network performance with time based on self-improving intelligent algorithms. Using AI in networking industries, and the increasing demand and scalability of networks makes it possible for engineers may be able to build and innovate more efficient and capable networking technologies.

AI will also help in managing a network, operations, and maintenance, as well as protecting the network beyond simply the scope of designing the network. Network monitoring is done with the help of algorithms that continuously keep track of anomalous traffic buildups to look for any malicious activities, distributed denial-of-service attacks, and hacking attempts. Supposedly, with the powerful algorithms of AI and ML, the network security is assumed to be at very high level. The reason being that AI will make network more intelligent enough to make itself secured, robust, dynamic, and agile to quickly isolate faulty or compromised sections and do needful actions to protect itself. AI will facilitate a network to find faster ways of issue resolution and more advanced methods of anticipating threats to keep the network clean. The real-time traffic handling, traffic prediction, traffic distribution though available multipath in the networks to avoid any futuristic traffic chocking or any interruption will be managed in a better way with AI. In that way, it will be very helpful to the network operations team who monitors a network in 24×7 environment to make the network operational all the time and minimum or less impacting network downtime. Also, with better traffic prediction of what is supposed to be a big event, it can make enable network managers better prepared for big events with the best possible optimization of traffic through the network.15.3.1 Intent-Based Networking

Intent-based networking (IBN) is such a networking approach that uses AI and ML approaches to automate all the organizational tasks of infrastructure management according to the business intent. It provides smooth network management with automation that can be applied across the network in accomplishing a specific purpose or intent and that is why is called as intent-based networking. The network is intelligent enough to translate the intents into network policies and then deploy then with suitable configurations into the network. We can consider IBN as an evolution of modern networking toward moving in the direction of making intelligent networks. Features such as smooth and simplified operations, network agility or dynamicity, improved security, and advanced automations are some of the important features of IBN. An intent is something that can be considered as the desired outcomes or the business fulfillment from the network orchestration, the business intent or the service request from the network administrator, or business owners without any human intervention. It lets the network administrators and the business owners decide the outcomes desired from a network orchestration to help the network auto-provision, configuration along with full lifecycle management of the application services. With the help of AI/ML, the high-level plans and policies of a business intent, that is, in a declarative model is automatically implemented and enforced into the network.

As shown in Figure 15.3, we can see the important steps in the intent-based networking. It starts with the business need at top that codify the core of IBN in the form of intent of the network operator. Through the intent, the business expresses the service request to the network that may be in human language or may follow a more traditional interface that must be translated into network. Therefore, the translation of the intent is very important step here and can use natural language processing, or various forms of machine learning (ML) and machine reasoning. At this step, using of machine reasoning can leverage domain-specific knowledge about networking to determine the realization or a kind of feasibility check for the desired intent in the given network context. The outcome of this step is the set of the best-possible networking and security policies that may follow network automation and is ready with suitable configurations for implementations in the network. After that, the activation

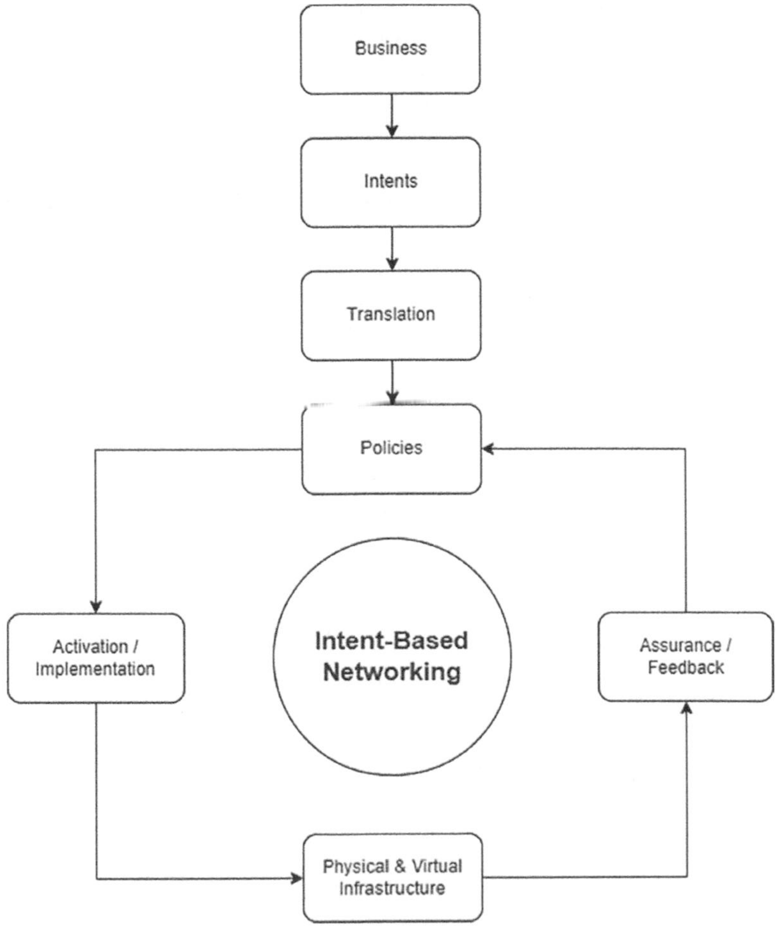

FIGURE 15.3 Important steps in intent-based networking.

step takes the network and security polices codified by the previous step and automates the policies across all of the network infrastructure elements including the virtual and physical, to optimize the performance, reliability, and security. Activation step may use different AI/ML techniques to accomplish the intended tasks, it has to determine about provisioning the network infrastructure for a desired quality of service (QoS) at each infrastructure element across the global network. Like providing the desired quality video along with ensuring other important network tasks are also operating as intended and at the same time handling real-time data with traffic optimization. The Assurance step as mentioned in the diagram is something that checks if the network is providing the service the intent calls for and the Activation step has implemented it correctly.

15.3.1 AI in Software-Defined Networking

Software-defined networking (SDN) is considered as an evolution in the modern networking technology that represents a promising networking architecture which combines central management and network programmability. It is used to separate the control plane from the data plane, where the network management is shifted to a central plane that placed with a network controller [8]. The controller provides a software controlled, programmable interface to the network administrator and represents the central brain of the network that leads to an advanced level of flexibility and network intelligence. Considering the recent trends, the research community has shown a keen interest in providing the SDN with the advanced algorithms of AI to have better learning and better decision-making abilities. Various subfields of AI namely machine learning, metaheuristics, and fuzzy inference systems are well considered as the research topics toward integrating SDN with AI. The involvement of AI/ML algorithms into networking has shown us ways to solve various networking problems [9] (Figure 15.4).

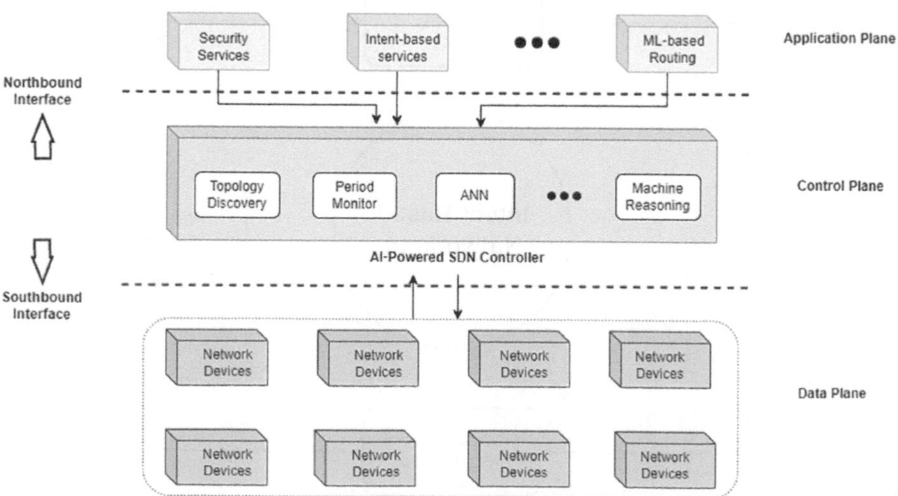

FIGURE 15.4 AI-powered software-defined networking.

Traditional networking problems such as routing, traffic classification, flow clustering, intrusion detection, load balancing, fault detection, quality of service (QoS), quality of experience optimization, admission control, and resource allocation are finding AI for suitable solution. However, with the evolution of SDN and the efforts made by industry and research community, the role of AI has been increased considerably. The combination of AI with SDN may lead to redesign the underlying network topologies and may drive toward more of network automation and can result in more profound network evolutions across all industries. Algorithms of machine learning can help the SDN controller to dynamically analyze the big heterogeneous data to perform network management, traffic engineering, network resources management, and programming the network dynamically [10]. At the same time, with the ML assistance, it would be easy for the SDN controller in continuous monitoring and collection of network configuration data, network state, information, and packet flow in real time. Such that the ML models can be trained based on historical network data for network performance optimization, data analysis, and automated network services provisioning intelligently.

15.3.1.1 Some Parameters for Selecting ML Algorithm to Power an SDN Controller

Traffic Classification: It enables network operators to control and allocate different services and plays an important role in network management, which is an important parameter to consider. Dynamic port-based approaches such as deep packet inspection and machine learning approaches are currently the most used traffic classification approaches. Port-based approaches have high classification accuracy but having limitations of application recognition based on pattern availability and high computational cost, also deep packet inspection is not able to identify the encrypted traffic flow. ML-based approaches are used to analyze encrypted traffic with a low computational cost as compared to traditional deep packet inspection approaches. Traffic classification can be easily done using ML algorithms by the collection of massive traffic flow data and knowledge extraction from that and thus ML applied to the SDN controller for the analysis of collected traffic data. ML approaches such as elephant flow-aware, QoS-aware, and application-aware are useful for traffic classification to make intelligent AI-powered SDN controllers.

Routing Optimization: A mathematical method for optimization is used to select the best routing decisions from some set of available choices. Routing decisions are very important for an SDN controller, and an optimized decision will help an SDN controller in managing the traffic flow in the best-possible way by modifying the flow tables of the devices in the data plane. ML methods such as reinforcement learning (RL) can be used for routing optimization using fast near-optimal routing solutions to help SDN controller for decision-making tasks. The SDN controller is considered as an agent and the network as the platform in RL-based routing optimization approach where the network and traffic states are considered in the state-space. Routing optimization using supervised learning approaches may consider network and traffic states as the input of a training dataset.

Traffic Prediction: Prediction of traffic is another important parameter useful for the SDN controller for the controlling actions and AI/ML approaches can be used for

TABLE 15.1
ML Methods Applied in SDN Controller

Supervised	Unsupervised	Reinforcement
K nearest neighbor	K means	Reinforcement learning (RL)
Decision Tree	Self-organizing map	Deep RL (DRL)
Neural Network (DNN)		RL game theory
Bayes' Theorem		
Hidden Markov Model		

the traffic prediction. ML methods are used for predictive modeling using regression models for prediction of likelihood of an outcome. Using machine learning modeling techniques, prediction algorithms can be used for the prediction of the outcome of undiscovered future events from current data to predict future problems. SDN controller can utilize the results of traffic prediction for efficient traffic routing decisions in advance and accordingly can distribute the policies dynamically into the data plane devices. SDN controller powered by ML techniques based on traffic prediction can avoid traffic congestion, improve QoS, and proactively monitor network health.

Network Security: Network security is another very important consideration while designing any network. SDN possess a new set of security challenges that appear as one of the biggest issues while designing SDNs. Certain threats to the controller may be due to the programming vulnerabilities, errors in configurations, or distributed denial-of-service attacks. AI and ML techniques are supposed to play a very significant role in securing SDN-based networks. The ANN-based approaches can be used in providing a collaborative intrusion prevention system to SDN such that each device will be responsible for collecting data to perform inputs for several ANNs. The ML methods are very supportive In order to secure the SDN controller or the SDN devices from any type of DDoS distributed denial-of-service attack, and avoid the SDN getting comprised any intrusion or any malicious activity for a hacking attempt. Self-organized map approach a variant of ANN based on unsupervised learning that can be used for detection of DDOS attack as a classification method to handle unlabeled input vector.

Table 15.1 lists machine learning algorithms that are used to SDN controller.

15.3.2 AI in the Telcom Networks

Currently, the deployment work of the fifth generation (5G) of wireless communication is in progress, and in the next decade it is supposed that beyond 5G (B5G) will be implemented. AI/ML techniques have the potential to solve the problems of 5G and B5G networks by analyzing the large amounts of data generated from the network. AI methods are supposed to touch different aspects in the designing and optimization of wireless network including channel measurements, modeling and estimation, physical-layer research, and network management and optimization [11]. AI will help in reducing manual efforts for the development, configuration, and management of the network along with providing better system performance, reliability, and adaptability of communication networks. AI is expected to be such a technology that will be rapidly used in B5G networks for better deployments and resolving certain network development issues. Some of the wireless network's development issues like channel and interference models are supposed to be very complicated due to dynamic nature of wireless communication channels in B5G scenarios. ML methods are widely expected to be very supportive in extracting unknown channel information automatically from the communication data and prior knowledge. In the B5G scenarios, with the continuous increase in the density of wireless access points, there is an urgent intervention needed in the global optimization of communication resources and the need for fine-tuning in the system settings.

Data-based analytics is nowadays very critical in the performance and operations of the telecommunication network, and service providers have worth regards for its importance, significantly corresponding to the advances in data analytics, including ML. There is always a need for predicting the change in user's demand on the network at a given point of time and that often requires efforts from the service provider in minimizing any negative quality of service effects. In order to meet this necessary demand from the user, the service provider will be required to reallocate some of its network resources. One approach to minimizing quality of service effects involves attempting to predict changes in user demand before they take place. Therefore, predictions from the user's change in demand will be very important for the service provider to initiate the necessary network changes. It would become extremely difficult for the service provider to improve the customer experience and increasing profit without any data for analytics and prediction. ML approaches now offer the ability to make more accurate predictions regarding changes in user demands to service providers. Further monitoring (system event logs, etc.) and operational data coming from the wide pool of resources in a heterogeneous network will be needed to give substantive insights on real-time networking processes. ML methods will assist in quick decision-making based on the big data analytics that is done on a wide pool of data coming from the resources that previously took slow human interactions, based on traditional network characteristics and optimization methods.

15.3.3 AI in Cyber Security

Cyber security has always been a concern and is very significant for any network, and the cyberattack surfaces are increasingly getting massive in modern enterprise environments. This indicates that the analysis and improvement in the cyber security posture needs more than simply the human intervention by doing a real-time analysis on a big data pool generated from alarms, events, logs, system health checks, and many more data resources. AI and ML are now getting popular as well in resolving the tedious cyber security challenge and have become very essential to information security [12]. The powerful techniques of AI and ML are now capable of analyzing very quickly the big data pool having millions of datasets to track a wide variety of cyber threats in real time. AI presents very suitable and supportive environment to offer cyber security in a fast-evolving cyberattacks with rapid multiplication of devices happening each day in the networks. AI is well capable of assisting in the cyber security challenge by identifying cyber threats and possible malicious activities where traditional software systems simply cannot keep pace with increasing new malware threats. ML models can be trained by using sophisticated algorithms for malware detection, can run pattern recognition, and detect any malicious activity before it enters into the system. Natural language processing using predictive intelligence can give threat intelligence on new anomalies and cyberattacks, and provide prevention strategies using the curation of data by scrapping through articles, news, and studies on cyber threats [13].

15.4 CONCLUSION

The increasing number of devices, equipment, and end points seeking connectivity; the increasing pool of device data; the development of the next-generation hardware technology; the evolving network and communication methods; and the increase in the number of people connected through networks (Internet) have made the network very complex. Managing a network is a key requirement, and considering the complex network it is getting day by day, we have to consider some cost-effective methods that are less prone to human errors. In that case, the AI is giving a software-centric approach that can perform a task on par with a human expert and hence is supposed to play a very critical role in taming the complexity of the growing complex IT networks [14].

There are certain benefits of having an AI/ML-based approach in networking that we have already seen in this chapter, although the methods of AI are already being used in networking and providing value to the industry. Certain examples are mentioned below to represent how beneficial the AI-driven approach would be in networking [15].

Anomalies Detection: With the help of AI, the detection of anomalies in the network devices can happen easily to help the network/security engineers for device management in the network. AI is very helpful to quickly detect violation of any time series anomalies in managing legacy devices, which are operational in today's network but were invented long ago and do not support current management messages.

Event Correlation: AI is very helpful in isolating and identifying network problems for faster resolution using various data-mining techniques that can quickly explore terabytes of data in a matter of minutes to provide event correlation and root cause analysis.

User experience Prediction: AI is supportive in dynamic planning and adjustment of bandwidth capacity and resources for better Internet performance based on user data usage prediction analysis by historical trends and current calendar information.

Self-driving: AI methods can allow the network and IT systems in self-correcting to give maximum uptime with prescriptive corrective actions to help fix problems and in that way will keep the network self-driving.

AI and ML are now the key components to information security that can swiftly analyze millions of datasets to track and cure a wide variety of cyber threats. AI is expected to bring various changes in the network paradigm and is going to play a vital role in futuristic network where the network is supposed to get smarter and adaptable to customer needs.

REFERENCES

1. Chemouil, P., Hui, P., Kellerer, W., Limam, N., Stadler, R., & Wen, Y. (2020). Guest editorial special issue on advances in artificial intelligence and machine learning for networking. *IEEE Journal on Selected Areas in Communications*, 38(10), 2229–2233.
2. Sivalingam, K. M. (2021). Applications of Artificial Intelligence, Machine Learning and related techniques for Computer Networking Systems. arXiv preprint arXiv:2105.15103.
3. Yao, G. (2016, December). Discussion on the application of artificial intelligence in computer network technology. In *International Conference on Electrical & Electronics Engineering and Computer Science*, Jinan, China.
4. Li, R., Zhao, Z., Zhou, X., Ding, G., Chen, Y., Wang, Z., & Zhang, H. (2017). Intelligent 5G: When cellular networks meet artificial intelligence. *IEEE Wireless Communications*, 24(5), 175–183.
5. McCarthy, J. (2004). What is artificial intelligence. Available: http://www-formal.stanford.edu/jmc/whatisai.html
6. Artificial intelligence and machine learning for networking and communications, Comsoc.org. Accessed: February 25, 2022. [Online]. Available: https://www.comsoc.org/publications/journals/ieee-jsac/cfp/artificial-intelligence-and-machine-learning-networking-and
7. What is artificial intelligence for networking?, Juniper Networks. Accessed: February 25, 2022. [Online]. Available: https://www.juniper.net/us/en/research-topics/what-is-ai-for-networking.html
8. Wu, Y. J., Hwang, P. C., Hwang, W. S., & Cheng, M. H. (2020). Artificial intelligence enabled routing in software defined networking. *Applied Sciences*, 10(18), 6564.
9. Latah, M., & Toker, L. (2016). Application of artificial intelligence to software defined networking: A survey. *Indian Journal of Science and Technology*, 9(44), 1–7.
10. Insight, A., The role of artificial Intelligence in Network Evolution, Analyticsinsight.net, 10-Feb-2022. Accessed: February 23, 2022. [Online]. Available: https://www.analyticsinsight.net/the-role-of-artificial-intelligence-in-network-evolution/
11. Wang, C. X., Di Renzo, M., Stanczak, S., Wang, S., & Larsson, E. G. (2020). Artificial intelligence enabled wireless networking for 5G and beyond: Recent advances and future challenges. *IEEE Wireless Communications*, 27(1), 16–23.
12. Dilek, S., Çakır, H., & Aydın, M. (2015). Applications of artificial intelligence techniques to combating cyber crimes: A review. arXiv preprint arXiv:1502.03552.
13. Veiga, A. P. (2018). Applications of artificial intelligence to network security. arXiv preprint arXiv:1803.09992.
14. Zeydan, E., & Turk, Y. (2020, May). Recent advances in intent-based networking: A survey. In *2020 IEEE 91st Vehicular Technology Conference (VTC2020-Spring)* (pp. 1–5). IEEE.
15. Apostolopoulos, J., Improving networks with artificial intelligence, Cisco Blogs, 05-Jun-2019. Accessed: February 25, 2022. [Online]. Available: https://blogs.cisco.com/networking/improving-networks-with-ai

Index